CELL APOPTOSIS AND CANCER

CELL APOPTOSIS AND CANCER

ALBINA W. TAYLOR
EDITOR

Nova Biomedical Books
New York

For permission to use material from this book please contact us:
Telephone 631-231-7269; Fax 631-231-8175
Web Site: http://www.novapublishers.com

LIBRARY OF CONGRESS CATALOGING-IN-PUBLICATION DATA

Cell apoptosis and cancer / Albina W. Taylor (editor).
p. ; cm.
Includes index.
ISBN-13: 978-1-60021-506-3
ISBN-10: 1-60021-506-8
1. Cancer. 2. Apoptosis. I. Taylor, Albina W.
[DNLM: 1. Neoplasms--etiology. 2. Apoptosis--physiology. QZ 202 C3924 2006]
RC254.6.C45 2006
616.99'4--dc22 2006036579

Published by Nova Science Publishers, Inc. ✣ New York

CONTENTS

PREFACE

Apoptosis is the regulated form of cell death. It is a complex process defined by a set of characteristic morphological and biochemical features that involves the active participation of affected cells in a self-destruction cascade. This programmed cell death plays a critical role in physiological functions such as cell deletion during embryonic development, balancing cell number in continuously renewing tissues and immune system development. Additionally, a dysregulation of apoptosis is underlying in numerous pathological situations such as Parkinson, Alzheimer´s disease and cancer. A number of studies have pointed out an association between consumption of fruits and vegetables, and certain beverages such as tea and wine, which are rich in polyphenols, with reduced risk of chronic diseases, including cancer. Apoptosis is also the regulatory mechanism involved in the removal of unnecessary cells during development and in tissue homeostasis in a wide range of organisms from insects to mammals. Cancer and cell apoptosis is the central focus of this significant new book.

Chapter I - The cancer metastatic process is thought to be complicated through a series of steps in which tumor cells migrate from primary site, penetrate into vasculature (intra-vasation), attach to endothelium and break the basement membrane (extra-vasation), migrate into extra cellular matrices, and eventually colonize and grow at distant site. During these steps, the disseminated cells from primary site must survive through mechanical stress like sheering force in the vasculature and attack from host immune system. Even when they survive in vasculature and attach to the endothelium and basement membrane, the intracellular signal for collagenase secretion for breakage of basement membrane, mobility activation for migration into extra cellular matrices, and proliferation stimuli must be trans-activated into the nucleus appropriately to form the metastatic loci. Whereas, recent reports indicated that only a few cell population of large numbers of cells disseminated from the primary site may gain the formation of metastatic loci. This is because a single tumor cell may get difficult to obtain every factor which is necessary in each step of metastatic process, thus the majority of cells might fail to complete some steps, and fall in apoptosis. This elimination of the majority of the cells and failure to form colonies has been termed "metastatic inefficiency". The regulation of apoptosis is influencing metastasis of tumor cells in the early stages after attachment to the endothelium, and considered to play an important role in this "metastatic inefficiency".

Recent studies supported the existence of cancer stem cells in solid tumors such as brain and breast tumors. The cancer stem cells possess the capability of self-renewal and may drive the autonomous tumor growth and formation of distant metastasis. The hypothesis here is that the cancer stem cells may form the clusters composed of different tumor cell progeny in which each express the one or some factors necessary for completion of metastatic process at the site of colonization, and successfully evade from the apoptotic elimination. Each cell in the cluster play each role which is dependent on its possessing factors such as destruction of basement membrane (activated collagenase), migration (mobility) etc. to conqur the every step for metastatic process as a cluster, then some of cells with stem cell characteristics in the cluster eventually form the metastatci loci, which would be termed "stem cell-mediated clustering metastasis theory".

Chapter II - Photodynamic therapy is a relatively new approach for the treatment of malignant and non-malignant diseases which is based on the action of photosensitizers, i.e. of molecules that have a cytotoxic effect when excited by light at appropriate wavelengths. The cytotoxic action of photosensitizers is due to their capability to convert light energy into oxidative processes leading to the production of new oxidizing chemical species (singlet oxygen, other reactive oxygen species or free radicals); these are highly unstable and can actively interact with the surrounding biomolecules, thus determining organelle damage and, eventually, cell death. Rose Bengal (RB) is a very efficient photosensitizer, whose photophysical properties are inactivated by addition of the quencher group acetate; this RB derivative (RB-Ac), even at low concentration, easily enters the cells where the original photoactive molecule is restored by specific esterases. It has already been demonstrated that, following green light irradiation, RB-Ac-treated cells exhibit a reduced viability, suggesting that the photoactive molecules are located in particularly sensitive cell sites. In fact, we report here, by fluorescence imaging techniques, that RB molecules redistribute dynamically through the cytoplasm, first in the endosomal compartment, then in the endoplasmic reticulum and the Golgi apparatus, and finally in the cytosol where different cell structures (including cytoskeleton components and mitochondria) undergo photodamage. Morphological evidence at light and electron microscopy as well as cytochemical and cytometric results showed that apoptotic cell death occurs in RB-Ac-photosensitized cells; interestingly, different apoptotic pathways are activated, as a consequence of the photodamage induced on different organelles.

Chapter III - In terms of an evolutionary biology of apoptosis pathways as cascade events it is significant that a full range of operative factors would implicate the dynamics of phenotypic variability characteristic of infiltrating neoplasia. It is towards the definition of such phenotypic variability that subsequent events prove or otherwise refute the concept of genetic instability as an operative framework in the development and potentiation of specific carcinogenetic pathways as predetermined biology of malignant lesions. It is particularly in view of a whole series of events borne out by the infiltration of tissues and spread of tumor cells as clumps of such cells that carcinogenesis would be purely definable as a predetermination of apoptotic versus anti-apoptotic events. These would operate in both initiation and subsequent characterization of lesion neogenesis as malignant disease. Only in such terms can there evolve a full series of events in the further definition of pathways as characteristic of both initiation and promotion of carcinogenetic agent action in further evolution of apoptosis versus anti-apoptosis transformation in the biology of neoplastic lesions.

Chapter IV - The ability to induce and execute apoptosis is essential for the maintenance and regulation of tissue and organ homeostasis. A common feature of malignant cells is their capability of evading apoptosis, which contributes to the high rate of resistance to radio- or chemotherapy of tumor cells. Although some progress has been made in inhibiting tumor growth by developing receptor-tyrosine-kinase antagonists that inhibit survival and growth promoting pathways, resistance against these and other compounds is mainly due to defects in apoptosis regulation. In contrast, low apoptotic thresholds may sensitize tumors to well tolerated and clinical established treatment regimens. Therefore, downstream targets of the apoptosis machinery, e.g. survivin, caspases or members of the bcl-2 family, seem to represent interesting new targets for cancer therapy. Modulating apoptotic thresholds can be achieved by reconstituting genetically lost or epigenetically silenced pro-apoptotic factors, e.g. by viral gene transfer or inducing re-expression with DNA methyltransferase or histone deacetylase inhibitors, or by inhibiting the activity or expression of anti-apoptotic factors. Here, RNAi-based approaches, e.g. against anti-apoptotic bcl-2, seem to hold great promise to overcome treatment resistance of a variety of malignant tumor types. Overall, restoring apoptosis sensitivity or lowering apoptotic thresholds will lead to better tolerated and more potent treatment strategies.

Chapter V - Tumors frequently contain activated $CD8^+$ T cells that have the potential to be cytolytic. However, spontaneous immune-mediated rejection of cancers occurs infrequently. Although there are several potential explanations for this observation, one leading mechanism may be the ultimate selection of tumor cells intrinsically resistant to immune-mediated apoptosis. Collectively, the integrated contribution of pro- and anti-apoptotic signaling factors in each individual tumor cell may decide its fate and ability to survive the attempts at killing by $CD8^+$ T cells. In this article, we will review how melanoma cells may be resistant to triggers of apoptosis. Overexpression of Bcl-2 family members, IAP family members, and survivin is commonly seen in melanoma. Resistance to cytotoxic granule-mediated lysis may be mediated through expression of Serpins, and escape from FasL-induced apoptosis may occur through downregulation of Fas and alterations in Fas signaling components. Finally, recent observations regarding the oncogenic events in melanomagenesis, such as activation of Notch signaling, have provided new insights into the potential biochemical regulation of an anti-apoptotic program.

Chapter VI - L-ascorbic acid (LAA) at concentrations of 0.25–1.0mM induced a dose- and time-dependent inhibition of proliferation in acute myeloid leukemia (AML) cell lines and also in leukemic cells from peripheral blood specimens obtained from three patients with AML. LAA induced oxidation of glutathione to oxidized form (GSSG) and subsequent H_2O_2 accumulation in a concentration-dependent manner, in parallel to induction of apoptosis. Induction of apoptosis in LAA-treated AML cells involved a dose-dependent increase of Bax protein, release of cytochrome C from mitochondria to cytosol, activation of caspase 9 and caspase 3, and cleavage of poly [ADP-ribose] polymerase. LAA treatment of cells induces a dose-dependent phosphorylation of extracellular signalregulated kinases (ERK) and results in activation of its catalytic domain. The small G protein Raf1 and MAPKactivated protein kinase 2 are activated by LAA as an upstream and a downstream regulator of ERK, respectively. Although the ERK pathway has been known to activate cell proliferation, pharmacologic inhibition of ERK reduces LAA-dependent apoptosis and growth inhibitory response of acute myeloid leukemia cell lines, suggesting that this signaling cascade positively regulates induction of apoptotic response by LAA.

Arsenic trioxide (As_2O_3) could induce *in vitro* growth inhibition and apoptosis of AML cells. As_2O_3 or LAA suppressed DNA-binding activity of NF-κB composed of p65/p50 heterodimer through preventing the degradation of IκB-α and the nuclear translocation of p65 subsequently as well as interrupting the binding of NF-κB with their consensus sequences. Inhibitory effect of As_2O_3 or LAA on NF-κB DNA activity was dependent upon intracellular glutathione (GSH) and H_2O_2 level, but not superoxide anion.

The addition of LAA to cells resulted in an immediate increase in intracellular total GSH content and GST activity accompanied by the uptake of cysteine. These results suggest a new role of LAA in intracellular sulfur condition and glutathione regulation. Early protein targets of LAA in leukemia cells were identified using a differential proteomics approach in NB4 human leukemia cells treated with LAA. A subunit of protein-disulfide isomerase (a thiol/disulfide exchange catalyst) and immunoglobulin-heavy-chain binding protein (BiP, identical to Hsp70 chaperone) showed quantitative expression profile differences. A myeloid leukemia associated antigen protein (a tropomyosin isoform) showed changes in pI as a result of phosphorylation. The addition of LAA to cells results in an immediate change in the intracellular thiol/disulfide condition and modulates disulfide bond formation in proteins.

Chapter VII - Adult T-cell leukemia (ATL) is a fatal malignancy of T lymphocytes caused by infection with human T-cell leukemia virus type I (HTLV-I) and remains incurable. Leukemic cells often evade apoptosis by expressing several anti-apoptotic proteins, down-regulation and mutation of proapoptotic genes, and alterations in signaling pathways that give them survival advantage and thereby allow them to resist therapy-induced apoptosis. Multiple oncogenic signals such as NF-κB, AP-1, phosphatidylinositol 3-kinase (PI3K)-Akt, and Janus kinase (JAK)-signal transducer and activator of transcription (STAT) signaling pathways are constitutively activated in HTLV-I-infected T cells, suggesting possible involvement of these signaling pathways in anti-apoptosis and leukemogenesis of ATL. Curcumin (diferuloylmethane), a natural yellow compound present in turmeric (*Curcuma longa* Linn), has been studied vigorously as a potent chemopreventive agent for cancer because of its inhibitory effect on proliferation and induction of apoptosis in several tumor cell lines. Curcumin prevents the growth of HTLV-I-infected T-cell lines and primary ATL cells by inducing apoptosis. Curcumin also inhibits the growth of HTLV-I-infected T-cell tumors implanted subcutaneously in SCID mice. These effects of curcumin are mediated by inhibiting the NF-κB, AP-1, PI3K-Akt, and JAK-STAT signaling pathways. The ability of curcumin to alter gene transcription and induce apoptosis in preclinical models is likely to be of particular relevance to chemoprevention and chemotherapy in ATL patients. This review describes the molecular mechanisms of curcumin-induced apoptosis in HTLV-I-infected T-cell lines and primary ATL cells.

Chapter VIII - The idea of treating cancer cells by means of hyperthermia dates back at least 5000 years. According to the Edwin Smith surgical papyrus a fire drill was used by "surgeons" to cauterize carcinoma in the breast. Over the past 30 years there are significant improvements in this "war against cancer". Despite this, the mortality of patients receiving treatments that die of or with their cancer is relatively high. The failure to achieve a reduction in mortality from some forms of cancer using conventional treatments, namely surgery, radiotherapy and chemotherapy has added a new dimension of interest in the potential of hyperthermia, both as treatment in its own right and in combination with other therapies.

The main reason why there has been so little interest for hyperthermia in modern cancer research is that the technologies used in the past cannot deliver effective and homogeneous heating of all sites, particularly the deeper seeded tumors. Also research and implementation has been hampered by lack of non-invasive temperature measurement. Recently, development of new techniques (nanotechnology, computer modelling, non-invasive thermometry) to control and direct the heat, stimulated the interest in hyperthermia again.

The objective of this artificially induced temperature rise is the treatment of tumors, directly by introducing irreversible biological damage or indirectly by enhancing the effects of other treatment regimes such as X-irradiation or chemotherapy in a synergistic way. Although the ultimate aim in cancer therapy is to kill cancer cells, the way the cells die is of importance. Nature employs two ways to kill cells; namely by necrosis and apoptosis. Necrosis usually induces an undesirable inflammatory response in vivo, while apoptotic death of cells is relatively free from this post-treatment problem. Apoptosis, characterized as programmed cell death, is considered a natural process in the homeostasis and defence systems of the body. Hyperthermia could be used in cancer therapy to induce apoptosis instead of necrosis to minimize adverse side effects. Although hyperthermia has been under investigation for decades, the precise mechanisms how hyperthermia induces apoptosis remains unclear. In order to further increase the effectiveness of hyperthermia the cellular mechanisms must be elucidated. Here, we outline the effects of hyperthermia both on cellular and molecular levels. Furthermore, several ways to induce hyperthermia and novel therapeutic strategies will be discussed.

Chapter IX- Programmed cell death or apoptosis is a controlled physiological process that has become a focus for cancer therapy. The caspase cascade system plays vital roles in the initiation, transduction and amplification of intracellular apoptotic signals. Caspases exist as inactive zymogens in cells and undergo a cascade of catalytic activation at the onset of apoptosis. However, new evidence points to an emergent role of alternative caspase-independent cell death pathways. Many tumor cells can unexpectedly survive the activation of caspases; thus, caspase-independent cell death programs are gaining increasing interest in cancer research. Tumor invasion and metastasis are associated with altered lysosomal trafficking and increased expression of lysosomal cathepsin proteases. Lysosome stability and lysosomal cathepsins have recently been suggested to play an important role in the early stages of apoptosis induced by various agents. During lysosomal rupture, cathepsin B and D enzymes are released from the lysosomal compartment to the cytosol where they contribute to the apoptotic cascade upstream of mitochondria. Such role of lysosomes and lysosomal cathepsins in initiation and execution of the apoptotic program has become clear in several models, and the existence of a lysosomal pathway of apoptosis is now generally accepted. We have explored the role of cathepsins as upstream signals contributing to apoptotic cell death. Biochemical and mechanistic studies have led to important insights in the underlying molecular mechanisms of cathepsin regulation. Here, we highlight recent advancements in understanding the molecular mechanism of cathepsin-mediated cell death, and outline possible implications for the development of novel cancer therapeutics that target tumor cell lysosomes.

EXPERT COMMENTARY

In: Cell Apoptosis and Cancer
Editor: Albina W. Taylor, pp. 3-4

ISBN: 978-1-60021-506-3

Commentary

ANTIAPOPTOTIC ACTIVITY OF HEAT SHOCK PROTEINS IS A CHALLENGE FOR CANCER TREATMENT

Alexander E. Kabakov
Department of Radiation Biochemistry, Medical Radiology Research Center, Obninsk, Russia.

The constitutively enhanced heat shock protein (Hsp) expression is typical for many human tumors and associated with their high resistance to therapy and poor prognosis. Moreover, some kinds of cancer treatment (e.g. hyperthermia, certain remedies of chemotherapy) can stimulate the cytoprotective Hsp induction in the treated tumor cells thereby lowering the expected therapeutic effect. In fact, up-regulation and active functioning of Hsp70, Hsp27 and Hsp90 in the invasive malignant cells can protect the latter from apoptosis following cytotoxic (therapeutic) exposures and, therefore, decrease the efficiency of chemo- and radiotherapy. In this respect, it seems of paramount importance to develop clinically applicable ways for down-regulating the expression and antiapoptotic activity of Hsps in patients' tumors.

In spite of some encouraging data obtained on model systems, at present, there are not appropriate drugs to be used in a clinical practice for down-regulating the *in vivo* Hsp expression. Although several compounds (Quercetin, KNK437, Emunin, NZ28) are hopefully described as inhibitors of the Hsp induction in stressed or drug-treated tumor cells, these compounds cannot yet be adopted in a clinical setting. So, the problem of pharmacological down-regulating Hsps in patients' tumors remains to be resolved.

With regard to the down-regulators of antiapoptotic activity of Hsps, two low-toxic derivatives of Geldanamycin, 17AAG and 17 DMAG, are potent inhibitors of the chaperone activity of Hsp90 and already used in clinics to induce or enhance apoptosis in the treated tumors. The pro-apoptotic action of these drugs is based on blocking the Hsp90-dependent activation of certain tumor growth-associated proteins and signaling pathways aimed at the cell survival, recovery and progression. Consequently, such apoptosis-promoting Hsp90 inhibitors can be synergistic with chemo- and radiotherapy by sensitizing the target tumor cells to cytotoxic agents including irradiation. It should be noted that the pharmacological inhibition of Hsp90 causes undesirable induction of cytoprotective Hsps (in particular, Hsp70 and Hsp27) in the treated tumor cells so that these additionally induced Hsps can impair the antitumor effects of chemo- and radiotherapy. Apparently, the clinically applicable blockers of the Hsp induction are herein required to be administered in combination with the Hsp90 inhibitors.

Unfortunately, no drugs specifically inhibiting the antiapoptotic activity of Hsp70 and Hsp27 have been developed so far. Creation of such drugs is the important task for modern pharmacology aimed at cancer treatment. A promising approach was described by C. Garrido and co-workers who successfully neutralized the tumor-associated antiapoptotic function of Hsp70 by means of a cell-permeable and Hsp70-binding peptide derived from the mitochondrial apoptosis-inducing factor (AIF). Perhaps, analogous approaches with use of Hsp-binding (and neutralizing) peptides derived from client proteins or cofactors of Hsp70 and Hsp27 will help to sequester the excess Hsp pool inside the tumor cells to overcome their resistance to chemo- and radiotherapy. Discovery of the other (non-peptide) inhibitors of Hsp70 and Hsp27 seems also possible.

In: Cell Apoptosis and Cancer
Editor: Albina W. Taylor, pp. 5-6

ISBN: 978-1-60021-506-3

Commentary

Development of Pro-Apoptotic Agents for Cancer Therapy

T. S. Sridhar

Department of Chemistry and Biochemistry, University of Windsor ON, Canada.

Biomarker is the new buzz word in clinical drug discovery and development. The poster child for this movement has been the Iressa story. The tyrosine kinase inhibitor gefitinib (Iressa) failed to demonstrate clinically significant benefit in phase III clinical trials after remarkable efficacy during Phase II trials in the treatment of lung cancer. The subsequent discovery that it may work preferentially in the subgroup with mutations in the EGF receptor has highlighted the importance of identification and utilization of appropriate surrogate markers of efficacy early in the course of clinical development to prevent expensive failures at late stages. Drug developers working on agents that enhance apoptosis of cancerous cells need to take note of this development and think seriously of identifying appropriate biomarkers of clinical responsiveness.

Three critical points in the cascade of apoptosis have been the prime targets of these investigational agents; a) agonist ligands for the death receptor, b) inhibitors of the BCL-2 family of anti-apoptotic proteins such as BCL-2 itself and BCL-XL, and c) inhibitors of intracellular anti-apoptotic proteins such as XIAP. While a number of molecules are in various stages of pre-clinical and clinical development there has been only one USFDA approved drug for cancer that works by targeting an apoptosis component, Bortezomib, a proteasomal blocker.

Aside from the experimental method of primary culture of tumor followed by treatment of the cultures ex-vivo with a battery of chemotherapeutics to detmine efficacy, there are few if any reliable methods of determining the responsiveness of a tumour to a specific agent. While it is clear that a variety of molecular alterations underlie non-responsiveness to a particular class of drugs, blockade of apoptosis is thought to be amongst the more important ones. Work on cancer cell lines has shown that different lines vary considerably in their susceptibility to the induction of apoptosis, and half a dozen different mechanisms are responsible for the blockade of apoptosis.

It may be instructive to analyze the activities of the critical components of apoptosis and the status of the different pathways of apoptosis in freshly obtained clinical samples. These data may permit the sub-grouping of tumours based on the mechanism of apoptosis evasion. This information would direct the choice of a particular pro-apoptotic agent with the highest likely hood of success in that situation.

Another way forward may be the design of trials where the pro-apoptotic investigational agents are used in combination with approved cytotoxic chemotherapeutic agents that induce apoptosis. The un-blocking of the apoptotic mechanism might permit the lowering of the dose of the chemotherapeutic agent and hence limit dose related toxicity, a primary limitation of current anti-cancer drugs.

The route from the bench to the bedside is littered with failures. Apoptotic drugs that are effective in cancer might have a better chance if we use our knowledge of the basic biology to sub-stratify patients.

RESEARCH ARTICLES

In: Cell Apoptosis and Cancer
Editor: Albina W. Taylor, pp. 9-21

ISBN: 978-1-60021-506-3

Chapter I

CANCER STEM CELL AND CLUSTERING METASTASIS THEORY; A HYPOTHESIS FOR EVASION OF APOPTOSIS IN CANCER METASTATIC PROCESS

Kanya Honoki[1,*], Hiromasa Fujii[1], Toshifumi Tsujiuchi[2]

[1]Department of Orthopedic Surgery, Nara Medical University, Japan;
[2]Department of Life Science, Faculty of Science and Technologies, Kinki University, Japan.

ABSTRACT

The cancer metastatic process is thought to be complicated through a series of steps in which tumor cells migrate from primary site, penetrate into vasculature (intra-vasation), attach to endothelium and break the basement membrane (extra-vasation), migrate into extra cellular matrices, and eventually colonize and grow at distant site. During these steps, the disseminated cells from primary site must survive through mechanical stress like sheering force in the vasculature and attack from host immune system. Even when they survive in vasculature and attach to the endothelium and basement membrane, the intracellular signal for collagenase secretion for breakage of basement membrane, mobility activation for migration into extra cellular matrices, and proliferation stimuli must be trans-activated into the nucleus appropriately to form the metastatic loci. Whereas, recent reports indicated that only a few cell population of large numbers of cells disseminated from the primary site may gain the formation of metastatic loci. This is because a single tumor cell may get difficult to obtain every factor which is necessary in each step of metastatic process, thus the majority of cells might fail to complete some steps, and fall in apoptosis. This elimination of the majority of the cells

* Correspondence concerning this article should be addressed to Kanya Honoki, MD, PhD. Department of Orthopedic Surgery, Nara Medical University, 840 Shijo-cho, Kashihara, NARA 634-8521, Japan.Tel: +81-744-22-3051, X2324, Fax: +81-744-25-6449, E-mail: kahonoki@naramed-u.ac.jp.

and failure to form colonies has been termed “metastatic inefficiency”. The regulation of apoptosis is influencing metastasis of tumor cells in the early stages after attachment to the endothelium, and considered to play an important role in this “metastatic inefficiency”.

Recent studies supported the existence of cancer stem cells in solid tumors such as brain and breast tumors. The cancer stem cells possess the capability of self-renewal and may drive the autonomous tumor growth and formation of distant metastasis. The hypothesis here is that the cancer stem cells may form the clusters composed of different tumor cell progeny in which each express the one or some factors necessary for completion of metastatic process at the site of colonization, and successfully evade from the apoptotic elimination. Each cell in the cluster play each role which is dependent on its possessing factors such as destruction of basement membrane (activated collagenase), migration (mobility) etc. to conqur the every step for metastatic process as a cluster, then some of cells with stem cell characteristics in the cluster eventually form the metastatci loci, which would be termed “stem cell-mediated clustering metastasis theory”.

Keywords: cancer stem cell, apoptosis, metastasis

Introduction

Metastasis, defined as tumor formation in secondary organs by the disseminated cancer cells from primary site, is one of the most major causes of death for cancer patients. The process of cancer metastasis is thought to be very complicated with multiple sequential steps including escape from primary tumor, intra-vasation (penetrate into vasculature system), survival in the circulation with avoidance of attack by host immune system, arrest at new sites and extra-vasation (breakage of the basement membrane and migrate into extracellular matrices), colonization and formation of tumor mass in new circumstances. Since the failure at any stage will result in failure of formation of metastatic loci, the process is considered to be “all or nothing”. Because of the requirement of specific capabilities to complete these steps, very few of the cells escaping from the primary tumor might survive to form the secondary tumor masses. This elimination of tumor cells is often termed “metastatic inefficiency” [1], and Fidler, I.J. depicted these processes as the “decathlon champion” model [2].

Apoptosis is defined as the “programmed cell death” which is a normal biological process to maintain homeostasis and control proper growth of organisms [3]. Cell cycle control and apoptosis regulation closely link to cellular proliferation, differentiation, senescence and cell death [4]. Many factors are involved in these separated end-point events such as cyclin dependent kinase (CDK) for proliferation, caspase activation for apoptosis, tissue function genes expression for differentiation. However, all of those are basically regulated in the response to the growth factors and their signals which linked each other by upstream signal transduction pathways [5].

Hanahan and Weinberg indicated that self-sufficiency in growth signals, insensitivity to antigrowth signals, evasion of apoptosis, limitless replicative potential, sustained angiogenesis, and tissue invasion and metastasis are hallmarks of cancer [6]. Establishment of metastasis depends on the characteristic abilities of invasion- angiogenic potential, insensitivity to growth inhibition, and lack of apoptosis [7].

Apoptosis is one of the most important key components of "metastatic inefficiency". Mechanical stresses like sheering forces in the vasculature have been suggested to lead to apoptosis in susceptible cells [8,9], and host immune system also may attack to the circulated tumor cells. Then, majority of the cells may encounter the lack of survival signals which would be necessary to survive; otherwise cells might result in apoptosis [10].

Once cells attach to the endothelium, extracellular matrix components, or basement membrane, the signals would be provided and cells might get the characteristics for survival at the places where are apart from the primary lesions. However, several studies indicated that more than 85% of the cells initially arrested in the lungs were lost [11,12]. Wong, *et al.* directly demonstrated that tumor cells died in the lungs after introduction into the circulation through apoptotic processes [13]. The regulation of apoptosis is considered to be essential for survival of metastatic tumor cells in the early stages after attachment to the pulmonary endothelium.

It seems to be a very difficult task to successfully establish the metastatic colony originated from a single cell at secondary site, because the cell needs to obtain every each factor and characteristics as mentioned above without any failures. Revisit to the concept of "embryonal rest" originated from back in 19th century give rise to the thought that cancer may arise from "embryonal rests" and rely on "cancer stem cells" for its autonomic growth [14,15]. A first report of the presence of cancer stem cells has made in acute myelogenous leukemia patients [16]. Then, subsequent reports demonstrated the evidence of cancer stem cells in the solid tumors, like breast [17], brain tumors [18], and even sarcomas [19]. The solid tumor is considered to have mixed; heterogeneous populations consisted of different phenotype of cancer cells. This heterogeneity, e.g. different phenotype of tumor cell population in a tumor, could be originated from sequential genetic mutations, or divergent cell proliferation/ differentiation from tumorigenic stem cells.

The heterogeneity would be essential for the establishment of malignant behavior including metastatic ability. The hypothesis of "cancer stem cell-mediated clustering metastasis theory" could explain how tumor cells might overcome the disadvantages encountering in the metastatic process, especially evasion from apoptotic stress.

Metastatic Process

The metastatic process consists of a series of steps (Figure 1), all of which must be successfully completed to give rise to a metastatic tumor. Many factors may contribute to complete each step and tumor cells may have to have the potential of expressing these factors.

Invasion and Intravasation

The first step of metastasis would be an escape from a primary tumor and invasion-intravasation into the circulatory blood system. Several factors might be involved in this step. Angiogenesis could provide an escape route by which cells can leave the tumor and enter into the vascular system. A variety of tumor cells are expressing elevated vascular endothelial growth factor (VEGF), a potent angiogenic factor, through several stimuli such as hypoxia [20]. Integrins, the adhesion molecules, play a primary role in mediating cellular adhesion to and interaction with molecules of extracellular matrix (ECM), such as collagen, fibrinogen, laminin, and vitronectin. Integrin-mediated signaling modulates the cell motility through the regulation of actin cytoskeleton [21,22]. Cell motility can play a role in early escape of the cells from the primary tumor. Integrins also controls both matrix metalloproteinases (MMPs) and serine proteinases, which may contribute to the invasion and intravasation [23]. Cadherins are another molecules that mediate cell-cell interactions and are integral components of adherens junctions and desmosomes, foaming the cytoplasmic cell-adhesion complex [24].

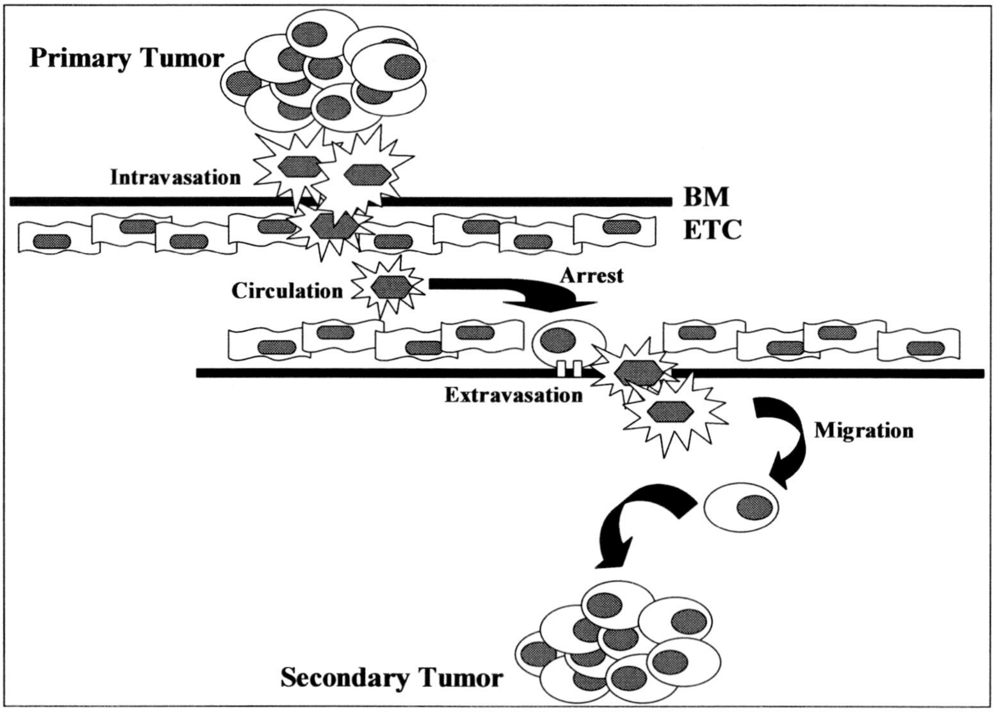

Figure 1. A classical model of metastasis. Cancer cells will acquire all of the factors which will be needed to complete the steps of metastasis, such as intravasation, arrest(adhesion), extravasation, migration and proliferation. BM: basement membrane, ETC: endothelial cells.

Transportation in the Circulation and Cell Arrest

This step would be the first step that cancer cells may encounter the physical and molecular stress in the circulation after escape from a primary tumor. This includes the hemodynamic forces and host immune surveillance. Many cells require adhesive contact with the ECM, and loss of integrin-ligand interactions can lead to downstream signaling resulting in cell death [25,26]. However, several recent studies suggested that there would be some initial loss of cells in the circulation in up to 20%, but that majority of cells are not lost to these early steps of the metastatic process [27-29]. This is in part through differential expression of cell surface adhesion molecules, and in part by development of "anchorage independent" phenotype in some tumor cells. Adhesion molecules like integrins are involved in the cell arrest interacting with collagen, fibrinogen and laminin in basement membrane and ECM.

Extravasation and Migration

When cells arrest on the basement membrane, signaling through adhesion molecules are believed to activate the cell motility and factors such as MMPs [21-23]. These activated cellular motility and MMPs may contribute to the process of extravasation as well as intravasation and migration. MMPs, of which 24 members are currently known, have the capacity to degrade all components of the basement membranes and extracellular matrix. Activation of MMPs is complex, requiring co-ordination among distinct families of intracellular, cell surface and extracellular proteinases [30,31]. Tissue inhibitor of metalloproteinases (TIMPs), endogenous inhibitor of MMPs, which regulate MMP activity are the major important factor to control a complex interaction between tumor cells and host cells. A fine balance between MMPs and TIMPs exists under normal conditions and this balance is often disturbed in the tumor environment. Both of these factors play a role in invasion, but also during the late stages of the metastatic process through their ability to mediate the interaction between tumor cells, the extracellular matrix, and stromal cells [32,33].

Secondary Tumor Growth and Formation of Metastatic Locus

The interaction between MMPs and TIMPs could be a large part of the cells' ability to survive and grow in their new environment as described above. Another major requirement factor which contributes to the progressive metastatic growth would be anoigogenesis [34]. Metastatic progression from micrometastasis to macroscopic metastases appears to be dependent on two factors: the growth of cancer cell itself and blood supply formation via the "angiogenic switch". Many factors such as chemokine ligands and their receptors (CXCL12-CXCR4, CCL21-CCR7), growth factors (VEGF, EGF, FGF etc.) and signaling molecules (Ras, p53 etc.) would control the cellular proliferation as well as angiogenesis to contribute to the growth of metastatic loci [7].

CANCER STEM CELL AND CARCINOGENESIS

Tissues in the body are considered to be derived from organ-specific stem cells that are defined by their capacity to undergo self-renewal and to differentiate into cell types that comprise each organ. Recent studies suggest that the properties of normal stem cells such as self-renewal, ability to differentiate, activation of anti-apoptotic pathways may be shared by cancer cells [35-37]. This supports the cancer stem cell hypothesis which indicates that tumors may contain and be driven by subpopulation of cellular components displaying stem cell properties, originally proposed about 150 years ago [38-40].

For most cancers, the target cell of transforming mutations is unknown. The "stochastic models" of carcinogenesis proposed that any cells may be the target of carcinogenesis, which would cause the transformation resulted from random mutation and subsequent clonal selection. In contrast, the "stem cell model" of carcinogenesis suggests that cancers originate in tissue stem or progenitor cells which are probably considered to be the initial targets of transformation. Dysregulation of self-renewal pathway initiates the expansion of the cell population leading further transformation with genetic and epigenetic changes which may result in cellular heterogeneity. Several experimental animal models have been affordable to assess of stem cell properties of tumor cell subpopulations. Dick *et al.* showed that the ability to transfer human leukemias into non-obese diabetic/severe combined immunodeficient (NOD/SCID) mice was retained in a small population of cancer cells that displayed the cell surface phenotype of stem cells, which comprised <1 in 10,000 leukemia cells [41]. Al-Hajj *et al.* showed that human breast cancer cells which expressed the stem cell characteristics, 1-10% of total cell population, are able to produce tumors implanted in NOD/SCID mice from as few as 200 of these cells [17]. In contrast, 2×10^4 cells which did not display the stem cell markers were unable to form tumors. Other groups extended and supported this cancer stem cell hypothesis in the same manner on human brain tumors [18,42], lung cancers [43], prostate cancers [44] and sarcomas [19]. These evidences support the hypothesis that only a few cancer cells are actually tumorigenic and that these tumorigenic cells could be considered as cancer stem cells [45].

Among the properties of normal stem cells, the most important and crucial property is that of self-renewal. All stem cells must self-renew and regulate the relative balance between self-renewal and differentiation to persist for the lifetime of the animal maintaining the normal organ function. From many observations suggesting the analogies between normal stem cells and tumorigenic cells, one could view a tumor as an abnormal organ initiated by a tumorigenic cancer cell which showed stem cell properties. Tissues, no matter normal or abnormal, are composed of heterogeneous combinations of cells, with different phenotypic characteristics and different proliferative potentials [46-49]. The tumorigenic cancer cells possibly could undergo processes that are analogous to the self-renewal and differentiation of normal stem cells, giving rise to phenotypically diverse progeny. This led to the presence of heterogeneous cell population in the most malignant tumors. Heterogeneity is a critical characteristic for malignant tumors. It may be essential for their ability of invasion and establishment of metastasis which need many factors to complete their process.

Apoptosis in Metastatic Process

It has been recognized that the metastatic process is an inherently inefficient for many years, indicating that only a small number of disseminated cancer cells can be detected at sites as a metastatic disease [50,51]. After arrest at a site, a cell must survive if it is going to progress to form a relevant metastatic disease.

Cancer cell destruction in the microcirculation and an inability to extravasate have been considered as the major contributors of metastatic inefficiency [52-54]. However, a group found that in the initial steps of the metastasis, survival in the circulation and survival of initial arrest were completed quite efficiently using B16 mice melanoma models [55,56]. They also found that initial survival of injected mammary carcinoma cells was found to be very high in liver with more than 80% [57]. These studies suggest that majority of cells (up to 80%) are not lost in the early steps of the metastatic process. In contrast, several studies demonstrated that intravenously injected tumor cells underwent apoptosis in the lungs of mice 24-48 hrs [13,58]. Furthermore, Luzzi et al. demonstrated that majority of the circulated cells (>80%) could survive and extravasate efficiently, however few extravasated cells (1/40) began to grow forming micrometastases, then much fewer micrometastases (1/100) could continue growth into macrometastatic tumors [56]. Therefore, the late steps of metastatic process subsequent to the extravasation could be the key components of the metastatic inefficiency.

Two events possibly involve in this majority of cell loss of the metastatic process, and these have to be conquered by tumor cells to complete the metastasis formation. Metastases have been known to sometimes occur decades after long period of "dormancy" [59,60]. Tumor dormancy is a well-established concept, referring "present alive but no gain of growth", and would be an important component of metastatic inefficiency. A large number of disseminated solitary cells and pre-angiogenic micrometastasis in secondary site was the major source of this dormancy [57,61]. Dormancy can arise from balanced proliferation and apoptosis within the tumor [61], and growth of metastatic tumor cells is regulated by a combination of factors inherent to individual cells and the microenvironment [62,63]. When tumor cells disseminated from primary site fail to take this balanced combination, they might fall in apoptosis after encountering several apoptotic stimuli, which would be immune regulation, Fas-TRAIL death receptor ligands signal [64,65], reactive oxygen species (NO, O_2^-, and H_2O_2) [66,67], hypoxia and such those activate apoptotic pathways. Thus, apoptosis of tumor cells is considered to be a significant barrier to establish the metastatic disease in any steps of arrest in circulation, interaction with endothelial cells, survival of solitary cells in secondary sites, initiation of micrometastatic growth, persistence of macrometastatic growth with angiogenesis. The ability of evasion from apoptotic stimuli is the most critical characteristics to establish the relevant metastatic disease for malignant tumors.

Cancer Stem Cell and Metastasis

Although there have been dramatic advances in the treatment of malignant tumors, the survival of patients with metastatic disease of most common solid tumors (lung, breast, prostate, colon cancers) and sarcomas has not significantly changed over several decades. This is possibly due to the ineffectiveness of current therapeutics to the cells in dormant state of metastatic lesions. In this scenario, another possible explanation of tumor dormancy might be maintained by existence of stem-like cells, which could self-renew themselves without differentiation and risk of apoptosis. Therefore, one reason for the failure of these treatments could be that current therapeutic modalities might fail to kill cancer stem cells effectively. The data of metastatic inefficiency accumulated above supports the hypothesis that most cancer cells in the dormant lesion lack the ability to form a new tumor at secondary site and fall into apoptosis, but only a small population of cells which possess the stem cell characteristics can lead to metastatic disease. Stem cells tend to be more resistant to chemotherapeutics and apoptosis than mature cell types [68]. If the same were true of cancer stem cells, then we could predict that these cells would be more resistant to chemotherapeutics and apoptosis and could establish the metastatic disease efficiently.

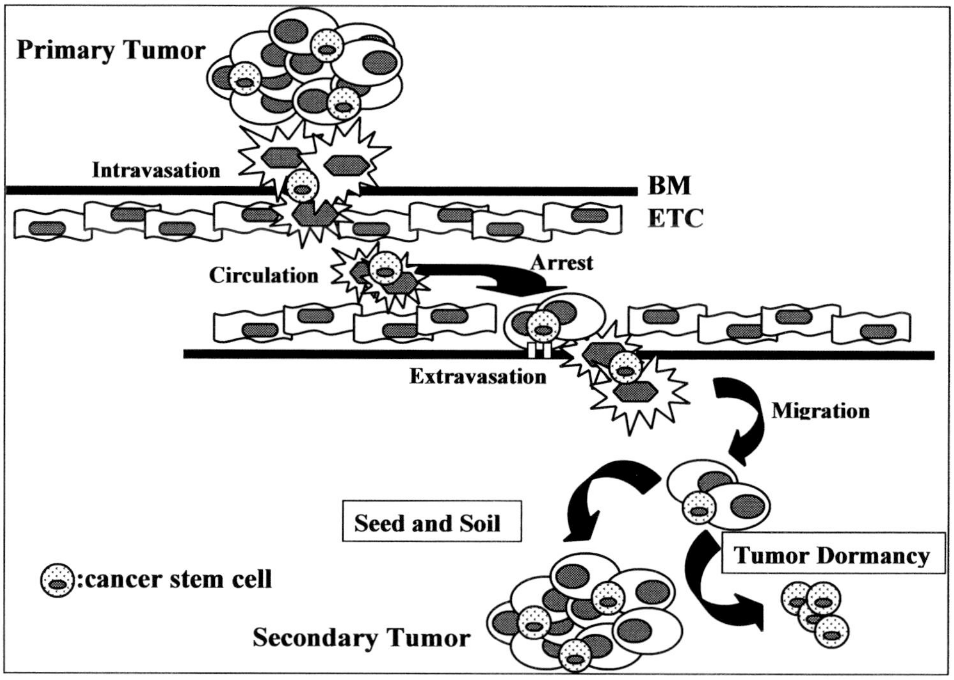

Figure 2. 'Clustering cancer stem cell' model of metastasis. The cancer stem cell may migrate into the tissues or vasculature system with the cells which may possess the ability to proceed the metastatic process such as matrix metalloproteinase, motility factors etc. The 'differentiated' tumor cells may give an appropriate microenvironment (niche) for cancer stem cells to allow the further growth and produce the metastatic lesions. BM: basement membrane, ETC: endothelial cells

It seems very difficult for single tumor cells to acquire every factor necessary for establishment of metastasis step by step in the process. Every single cancer cell may not be able to develop every characteristic which should be needed to complete the complex metastatic process, and cells are not able to continue to proliferate at secondary sites without an appropriate cellular microenvironment, postulated as the 'seed and soil' model back in 19th century [69,70]. Cancer stem cells may differentiate into the cells which can grow at the appropriate environment (niche) to establish the metastatic lesion. But, how could cancer stem cells which may not have any factors to be needed for metastasis migrate into the secondary site apart from primary tumor?

The hypothesis here is that tumor cells may not migrate as a single cell, but as a cluster which contains stem-like cells and other differentiated cell populations compensating each other for necessary factors to complete the complex metastatic process (Figure 2). One cell population in the cluster which expresses the adhesion molecules such as integrins may lead cell cluster to colonize the site of attachment. Then cells possessing the MMPs will degrade the extracellular matrix and basement membrane. After the completion of matrix degradation, cells which possess motility factors will migrate into the surrounding tissues. One population may have both adhesion molecules and MMPs or motility factors. There is a possibility that cancer stem cells may develop a new cluster at the site of colonization. Then, the cluster will become dormant state of micrometastasis, until it gets appropriate proliferation stimuli, otherwise differentiated cells other than stem-like cells in the cluster might fall into apoptosis. Only cancer stem cell in the dormant lesion can evade the apoptotic process, thus grow into relevant metastasis, when the microenvironment (niche) is appropriate for the cell. One possible explanation for this is that more differentiated tumor cells within the dormant lesion may produce factors that promote the proliferation and/or survival of cancer stem cells or regulate their stemness in a manner of niche for normal stem cells.

Clustering metastasis theory might be able to explain the phenomenon of both 'tumor dormancy' and 'seed and soil' organ specificity of metastasis which are related to the "metastatic inefficiency".

Conclusion

Both 'tumor dormancy' and 'seed and soil' model may be explained by the presence of stem-like cells in the metastatic process, and these cancer stem cells might be the essential factor for overcoming the "metastatic inefficiency". The targeting this stem-like cells may become a new strategy of treatments to conquer the life-threatening events for cancer patients.

References

[1] Weiss, L. (1990) Metastatic inefficiency. *Adv. Cancer Res.*, *54*, 159-211.

[2] Fidler, I.J. (1990) Critical factors in the biology of human cancer metastasis: twenty-eighth G.H.A. Clowes memorial award lecture. *Cancer Res.*, *50*, 6130-6138.

[3] Kerr, J.F., Winterford, C.M., Harmon, B.V. (1994) Apoptosis. Its significance in cancer and cancer therapy. *Cancer*, *73*, 2013-2026.

[4] Abrams, J.M. (2002) Competition and compensation: coupled to death in development and cancer. *Cell, 110*, 403-406.

[5] Blagosklonny, M.V. (2003) Apoptosis, proliferation, differentiation: in search of the order. *Semin. Cancer Biol.*, *13*, 97-105.

[6] Hanahan, D., Weinberg, R.A. (2001) The hallmarks of cancer. *Cell, 100*, 57-70.
[7] Chambers, A.F., Groom, A.C., MacDonald, I.C. (2002) Dissemination and growth of cancer cells in metastatic sites. *Nat Rev Cancer, 2*, 563-572.
[8] Weiss, L., Nannmark, U., Johansson, B.R., Bagge, U. (1992) Lethal deformation of cancer cells in the microcirculation: a potential rate regilator of hematogenous metastasis. *Int. J. Cancer, 50*, 103-107.
[9] Weiss, L., Elkin, G., Barbara-Guillem, E. (1993) The differential resistance of B16 wild-type and F10 cells to mechanical trauma *in vitro*. *Metastasis. 13*, 92-101.
[10] Frisch, S.M., Francis, H. (1994) Disruption of epithelial cell-matrix interactions induces apoptosis. *J. Cell Biol., 124*, 619-626.
[11] Fidler, I.J. (1970) Metastasis: quantitative analysis of distribution and fate of tumor emboli-labeled with125 I-5-iodo-2'-deoxyuridine. *J. Natl. Cancer Inst., 45*, 773-782.
[12] Fidler, I.J., Nicholson, G.L. (1977) Fate of recirculating B16 melanoma metastatic variant cells in parabiotic syngeneic recipients. *J. Natl. Cancer Inst., 58*, 1867-1872.
[13] Wong, C.W., Lee, A., Yu, L.S., Dong, Y. Kao, G., Al-Mehdi, A.B., Bernhard, E.J., Muschel, R.J. (2001) Apoptosis: An early event in metastatic inefficiency. *Cancer Res., 61*, 333-338.
[14] Oberling, C. (1952) The riddle of cancer. New Haven: Yale University Press, p.238
[15] Rather, L. (1978) The genesis of Cancer. Baltimore: Johns Hopkins University Press, p262
[16] Lapidot, T., Sirard, C., Vormoor, J., et al. (1994) A cell initiating human acute myeloid leukemia after transplantation into SCID mice. *Nature, 367*, 645-648.
[17] Al-Hajj, M., Wicha, M.S., Benito-Hernandez A., Morrison, S.J., Claeke, M.F. (2003) Prospective identification of tumorigenic breast cancer cells. *Proc. Natl. Acad. Sci. USA., 100*, 3983-3988.
[18] Singh, S.K., Hawkins, C., Clarke, I.D., et al. (2004) Identification of human brain tumour initiating cells. *Nature, 432*, 396-401.
[19] Gibbs, C.P., Kukekov, V.G., Reith, J.D., Tchigrinova, O., Suslov, O.N., Scott, E.W., Ghivizzani, S.C., Ignatova, T.N., Steindler, D.A. (2005) Stem-like cells in bone sarcomas: Implications for tumorigenesis. *Neoplasia, 7,* 967-976.
[20] Rofstad, E.K. and Danielsen, T. (1999) Hypoxia-induced metastasis of human melanoma cells: involvement of vascular endothelial growth factor-mediated angiogenesis. *Br. J. Cancer, 80*, 1697-1707.
[21] Juliano, R.L. (2002) Signal transduction by cell adhesion receptors and the cytoskeleton: functions of integrins, cadherins, selectins, and immunoglobulin-superfamily members. *Annu Rev Pharmacol Toxicol., 42,* 283-323.
[22] Price, L.S. and Collard, J.G. (2001) Regulation of the cytoskeleton by Rho-family GTPases: implications for tumour cell invasion. *Semin. Cancer Biol., 11*, 167-173.
[23] Giannelli, G., Astigiano, S., Anotonaci, S., Morini, M., Barbieri, O., Noonan, D.M., Albini, A. (2002) Role of the alpha3beta1 and alpha6beta4 integrins in tumor invasion. *Clin Exp Metastasis. 19,* 217-223.
[24] Cavallaro, U., Schaffhauser, B., Christofori, G. (2002) Cadherins and the tumour progression: is it all in a switch? *Cancer Lett., 176,* 123-128.
[25] Raff, M.C. (1992) Social controls on cell survival and cell death. Nature, *356,* 397-400.

[26] Stupack, D.G., Cheresh, D.A. (2002) Get a ligand, get a life: integrins, signaling and cell survival. *J Cell Sci.*, *115*, 3729-3738.

[27] Luzzi, K.J., MacDonald, I.C., Schmidt, E.E., Kerkvliet, N., Morris, V.L., Chambers, A.F., Groom, A.C. (1998) Multistep nature of metastatic inefficiency: dormancy of solitary cells after successful extravasation and limited survival of early micrometastases. *Am J Pathol.*, *153*, 865-873.

[28] Cameron, M.D., Schmidt, E.E., Kerkvliet, N., Nadkami, K.V., Morris, V.L., Groom, A.C., Chambers, A.F., MacDonald, I.C. (2000) Temporal progression of metastasis in lung: cell survival, dormancy, and location dependence of metastatic inefficiency. *Cancer Res.*, *60*, 2541-2546.

[29] Naumov, G.N., MacDonald, I.C., Weinmeister, P.M., Kerkvliet, N., Chambers, A.F. (2002) Persistence of solitary mammary carcinoma cells in a secondary site: a possible contributor to dormancy. *Cancer Res.* *62*, 2162-2168.

[30] Overall, C.M., Lopez-Otin, C. (2002) Strategies for MMP inhibition in cancer: innovations for the post-trial era. *Nat. Rev. Cancer, 2,* 657-672.

[31] Ellerbroek, S.M. Stack, M.S. (1999) Membrane associated matrix metalloproteinases in metastasis. *Bioessays*, 1999, 940-949.

[32] Chambers, A.F., Matrisian, L.M. (1997) Changing views of the role of matrix metalloproteinases in metastasis. *J. Natl. Cancer Inst.*, *89*, 1260-1270.

[33] McCawley, L.J., Matrisian, L.M. (2001) Matrix metalloproteinases: they're not just for matrix anymore! *Curr. Opin. Cell. Biol.*, *13*, 534-540.

[34] Folkman, J. (2002) Role of angiogenesis in tumor growth and metastasis. *Semin. Oncol.*, *29*, 15-18.

[35] Reynolds, B.A., Weiss, S. (1996) Clonal and population analyses demonstrate that an EGF-responsive mammalian embryonic CNS precursor is a stem cell. *Dev. Biol., 175*, 1-13.

[36] Jamieson, C., Ailles, L., Dylla, S., et al. (2004) Granulocyte-macrophage progenitors as candidate leukemic stem cells in blast-crisis CML. *N. Engl. J. Med.*, *351*, 657-667.

[37] Dontu, G., Al-Hajj, M., Abdallah, W.M., Clarke, M.F., Wicha, M.S. (2003) Stem cells in normal breast development and breast cancer. *Cell Prolif., 36 Suppl 1*, 59-72.

[38] Cohnheim J. (1867) Ueber entzundung und eiterung. *Path. Anat. Physiol. Kiln. Med., 40*, 1-79.

[39] Durante, F. (1874) Nesso fisio-pathologico tra la struttura dei nei materni e la genesi di alcuni tumori maligni. *Arch. Memor. Observ. Chir. Pract.*, *11*, 217-226.

[40] Cohnheim, J. (1875) Congenitales, quergestreiftes Muskelsakon der Nireren. *Virchows Arch., 65*, 64.

[41] Bonnet, D., Dick, J.E. (1997) Human acute myeloid leukemia is organized as a hierarchy that originates from a primitive hematopoietic cell. *Nat. Med.* *3*, 730-737.

[42] Singh, S.K., Clarke, I.D., Hide, T., Dirks, P.B. (2004) Cancer stem cells in nervous system tumors. *Oncogene*, *23*, 7267-7273.

[43] Kim, C.F., Jackson, E.L., Woolfenden, A.E., et al. (2005) Identification of bronchioalveolar stem cells in normal lung and lung cancer. *Cell*, *121*, 823-835.

[44] Maitland, N.J., Collins, A.T., Bryce, S., et al. (2005) Prospective identification of tumorigeneic prostate cancer stem cells. [abstract 2518] *Proceedings of the 96th Annual Meeting of the AACR*, pp 10946.

[45] Hamburger, A.W., Salmons, S.E. (1977) Primary bioassay of human tumor stem cells. *Science, 197*, 461-463.

[46] Fidler, J.J., Kripke, M.L. (1977) Metastasis results from preexisting variant cells within a malignant tumor. *Science, 197*, 893-895.

[47] Fidler, J.J., Hart, I.R. (1982) Biological diversity in metastatic neoplasms: origins and implications. *Science, 217*, 998-1003.

[48] Heppner, G.H. (1984) Tumor heterogeneity. *Cancer Res., 44*, 2259-2265.

[49] Nowell, P.C. (1986) Mechanisms of tumor progression. *Cancer Res., 46*, 2203-2207.

[50] Southam, C.M., Brunschwig, A. (1961) Quntitative studies of autotranplantation of human cenacer. *Cancer, 15*, 971-978.

[51] Salsbury, A.J. (1975) The significance of circulating cancer cell. *Cancer Treatment Rev., 2*, 55-72.

[52] Stetler-Stevenson, W.G., Aznavoorian, S., Liotta, L.A. (1993) Tumor cell interactions with estracelular matrix during invasion and metastasis. *Annu. Rev. Cell Biol., 9*, 541-543.

[53] Nicholson, G.L. (1991) Molecular mechanisms of cancer metastasis: tumor and host properties and the role of oncogenes and suppressor genes. *Curr. Opin. Oncol., 3*, 75-92.

[54] Liotta, L.A., Stetler-Stevenson, W.G. (1991) Tumor invasion and metastasis: an imbalance of positive and negative regulation. *Cancer Res., 51*, 5054s-5059s.

[55] Koop, S., Macdonald, I.C., Luzzi, K., Schmidt, E.E., Moris, V.L., Grattan, M., Khokha, R., Cambers, A.F., Groom, A.C. (1995) Fate of melanoma cells entering the microcirculation: over 80% survive and extravasate. *Cancer Res., 55*, 2520-2523.

[56] Luzzi, K.J., MaDonald, I.C., Schmidt, E.E., Kerkviliet, N., Morris, V.L., Chambers, A.F., Groom, A.C. (1998) Multistep nature of metastatic inefficiency. Dromancy of solitaly cells after successful extravasation and limited survival of early micrometastasis. *Am. J. Pathol. 153*, 865-873.

[57] Naumov, G.N., MacDonald, I.C., Weinmeister, P.M., Kerkvliet, N., Nadkarni, K.V., Wilson, S.M., Morris, V.L., Groom, A.C., Chambers, A.F. (2002) Persistence of solitary mammary carcinoma cells in a secondary site: a possible contributoe to dormancy. *Cancer Res., 62*, 2162-2168.

[58] Cameron, M.D., Schmidt, E.E., Kerkvilet, N., Nadkarni, K.V., Morris, V.L., Groom, A.C., Chambers, A.F., MacDonald, I.C. (2000) Temporal progression of metastasis in lung: Cell survival, dormancy, and location dependence of metastatic inefficiency. *Cancer Res., 60*, 2541-2546.

[59] Meltzer, A. (1990) Dormancy and breast cancer. *J. Surg. Oncol., 43*, 181-189.

[60] Karrison, T.G., Ferguson, D.J., Meier, P. (1999) Dormancy of mammary carcinoma after mastectomy. *J. Natl. Cancer Inst., 91*, 80-85.

[61] Holmgren, L., O'Reilly, M.S., Folkman, J. (1995) Dormancy of micrometastases: balanced proliferation and apoptosis in the presence of angiogenesis suppression. *Nature Med., 1*, 149-153.

[62] Nicolson, G.L. (1993) Paracrine and autocrine growth mechanisms in tumor metastasis to specific sites with particular emphasis on brain and lung metastasis. *Cancer Metastasis Rev.*, *12*, 325-343.

[63] Radinsky, R. (1995) Modulation of tumor cell gene expression and phenotype by the organ-specific metastatic environment. *Cancer Metastasis Rev.*, *14*, 323-338.

[64] Owen-Schaub, L.B., van Golen, K.L., Hill, L.L., Price, J.E. (1998) Fas and Fas ligand interactions suppress melanoma lung metastasis. *J. Exp. Med.*, *188*, 1717-1723.

[65] Takeda, K., Hayakawa, Y., Smyth, M.J., Kayagaki, N., et al. (2001) Involvement of tumor necrosis factor-related apoptosis-inducing ligand in surveillance of tumor metastasis by liver natural killer cells. *Nature Med.*, *7*, 94-100.

[66] Lala, P.K., Chakraborty, C. (2001) Role of nitric oxide in carcinogenesis and tumour progression *Lancet Oncol.*, *2*, 149-156.

[67] Al-Mehdi, A.B., Tozawa, K., Fisher, A.B., Shientag, L., Lee, A., Muschel, R.J. (2000) Intravascular origin of metastasis from the proliferation of endothelium-attached tumor cells: a new model for metastasis. *Nature Med.*, *6*, 100-102.

[68] Harrison, D.E., Lernaer, C.P. (1991) Most primitive hematopoietic stem cells are stimulated to cycle rapidly after treatment with 5-fluorouracil. Blood, *78*, 1237-1240.

[69] Paget, S. (1889) The distribution of secondary growth in cancer of the breast. *Lancet, 1*, 99-101.

[70] Ewing, J. (1928) in *Neoplastic Diseases. A Treatise on Tumors*. (W.B.Saunders Co., Philadelphia & London), pp 77-89.

In: Cell Apoptosis and Cancer
Editor: Albina W. Taylor, pp. 23-45
ISBN: 978-1-60021-506-3

Chapter II

APOPTOSIS AS A CONSEQUENCE OF MULTIPLE ORGANELLE DAMAGE IN CULTURED TUMOR CELLS AFTER PHOTOSENSITIZATION WITH ROSE BENGAL ACETATE

C. Soldani[1,*], M. G. Bottone[1,2], A. Fraschini[1], A. C. Croce[2], G. Bottiroli[2], A. I. Scovassi[2] and C. Pellicciari[1]

[1]Dipartimento di Biologia Animale (Laboratorio di Biologia Cellulare e Neurobiologia), University of Pavia; [2]Istituto di Genetica Molecolare del CNR, Pavia, Italy.

ABSTRACT

Photodynamic therapy is a relatively new approach for the treatment of malignant and non-malignant diseases which is based on the action of photosensitizers, i.e. of molecules that have a cytotoxic effect when excited by light at appropriate wavelengths. The cytotoxic action of photosensitizers is due to their capability to convert light energy into oxidative processes leading to the production of new oxidizing chemical species (singlet oxygen, other reactive oxygen species or free radicals); these are highly unstable and can actively interact with the surrounding biomolecules, thus determining organelle damage and, eventually, cell death. Rose Bengal (RB) is a very efficient photosensitizer, whose photophysical properties are inactivated by addition of the quencher group acetate; this RB derivative (RB-Ac), even at low concentration, easily enters the cells where the original photoactive molecule is restored by specific esterases. It has already been demonstrated that, following green light irradiation, RB-Ac-treated cells exhibit a reduced viability, suggesting that the photoactive molecules are located in particularly sensitive cell sites. In fact, we report here, by fluorescence imaging techniques, that RB molecules redistribute dynamically through the cytoplasm, first in the endosomal compartment, then in the endoplasmic reticulum and the Golgi apparatus, and finally in

* Correspondence concerning this article should be addressed to Cristiana Soldani, Laboratorio di Biologia Cellulare e Neurobiologia, Dipartimento di Biologia Animale – Università degli Studi di Pavia, Piazza Botta, 10; 27100 Pavia (Italy). Phone +39 0382 986420; Fax +39 0382 986325; E-mail: soldani@unipv.it

the cytosol where different cell structures (including cytoskeleton components and mitochondria) undergo photodamage. Morphological evidence at light and electron microscopy as well as cytochemical and cytometric results showed that apoptotic cell death occurs in RB-Ac-photosensitized cells; interestingly, different apoptotic pathways are activated, as a consequence of the photodamage induced on different organelles.

Keywords: Apoptosis, Rose Bengal acetate, Immunohistochemistry, Fluorescence microscopy, PDT

INTRODUCTION

Photodynamic Therapy

Over the last three decades, photodynamic therapy (PDT) has become a useful clinical tool and a viable alternative to other established therapies in the treatment of cancer and other diseases. In fact, PDT has potential application to a wide variety of cancers, including lung, cervical, esophageal, gastric, and breast carcinoma (Hopper, 2000; Dolmans et al., 2003). In dermatology, PDT is considered for the treatment of basal cell carcinoma, actinic keratoses, cutaneous lesions of Bowen's disease, psoriasis, cutaneous T-cell lymphoma, acne, and more recently it has been proposed for photo-rejuvenation (Ibbotson, 2002; Nestor et al., 2006). The advantages in this field include the capacity for non-invasive targeted therapy (*via* the topical application of the drug and local irradiation of involved areas) as well as the ability to generate excellent cosmetic results with minimal discomfort. In ophthalmology, the recent application of Verteporfin for the PDT of age-related macular degeneration (AMD) represented the first breakthrough in the use of PDT to treat an otherwise untreatable condition (Brown and Mellish, 2001).

Several disciplines have contributed to the development of PDT: chemistry, in the identification and preparation of new photosensitizing agents (photosensitizers, PS); biology, in the elucidation of the cellular processes involved in PDT; pharmacology and physiology, in identifying the mechanisms of distribution of PS in an organism; and last but not least physics, in the construction of suitable light sources, imaging devices and optical sensors, and in the development of dosimetric concepts and spectroscopic methods for determining PS concentrations in different cells and tissues. Biophysics have also helped to focus on the role of pH for PS accumulation, dose rate effects, oxygen depletion, temperature, and optical penetration of light at different wavelengths into various types of tissue: these are all important factors to be taken into account to optimize the effects of PDT (Peng, 2006).

Essentially, PDT involves the administration of a PS, which selectively accumulates in target cells, followed by exposure of the cells to a light source. Excitation of the PS molecules with light generates reacting species leading to oxidative damage of cellular structures. The photochemical reaction is described by the Jablonski diagram (Figure 1). The ground electronic state of the PS is a singlet state (S_0). On absorption of light of appropriate wavelengths (1), the PS is excited to the short-lived first excited singlet-state (S_1), from which it can return to the S_0-state by emitting the absorbed energy as fluorescence (2) or by internal conversion (3). Alternatively, the S1 PS can convert to the first excited triplet state (T_1) by intersystem crossing (4). The loss of energy from T_1 to the S_0-state can occur by emitting a phosphorescence photon (5). However, the T_1-state is sufficiently long-lived to favor the interaction of the excited PS with the surrounding molecules, with the generation of cytotoxic species (6). Therefore the photodynamic action is mostly mediated by the T_1-state.

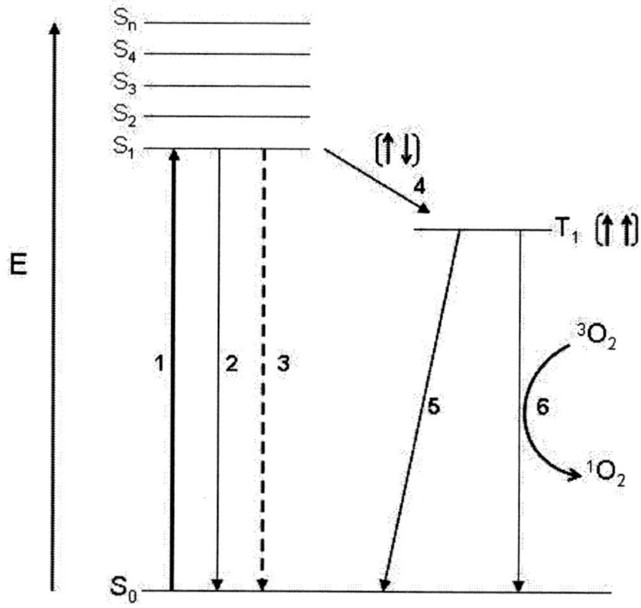

Figure 1. Modified Jablonski diagram. Photophysical processes: 1-absorption, 2-fluorescence, 3-internal conversion, 4-intersystem crossing, 5-phosphorescence and 6-formation of singlet oxygen $^1O_2$ by energy transfer from T1 photosensitizer to triplet oxygen $^3O_2$.

The production of reactive species can follow two mechanisms: Type I mechanisms involve the interaction of the excited state of the PS and a substrate, with the hydrogen- atom abstraction or electron transfer reactions to yield free radicals and radical ions, and can be prevalent in anoxic or hypoxic conditions; Type II mechanisms occur in oxygenated environments and consist in the direct interaction between the excited PS and oxygen, to generate singlet oxygen. Singlet oxygen is usually the predominant cytotoxic agent produced during PDT. The high reactivity of singlet oxygen and its consequently short lifetime (less than 40μs, in biological systems) makes the action of this chemical species restricted to the surrounding molecules, with an action radius of less that 0.02 μm (Moan and Berg, 1991).

The damage of cellular structures can possibly lead to cell death *via* necrosis or apoptosis (Dougherty et al., 1998). The way by which PDT drives cells into either pathway is largely dependent on the location of the PS inside the cell. This subcellular location is in turn governed by the PS chemical nature (i.e., molecular weight, lipophilicity, amphiphilicity, ionic charge and protein binding characteristics), as well as by the PS concentration, the incubation time, and the phenotype of the target cell.

Photosensitizing Agents

Among the several natural and synthetic dyes that can act as PS, the molecules useful for PDT applications need to respond to the following ideal characteristics: chemical purity with a constant composition, no toxicity in the absence of light, *in vivo* selectivity for the target lesion to be treated, high absorbance at wavelengths ensuring optimal penetration in biological tissues, rapid clearance from the body to avoid unwanted effects such as skin photosensitization. The PS should also be stable enough to avoid rapid degradation processes, although the photobleaching of the PS can be an advantage as well, because it can shorten the duration of generalized photosensitivity after PDT, thus increasing the treatment selectivity (Bonnet, 1999).

Three generations of PS can be distinguished. The first generation is made by hematoporphyrin derivatives. Porphyrins are an ubiquitous class of naturally occurring molecules, involved in various biological processes, such as oxygen transport, catalysis and pigmentation. The tetrapyrrolic ring, called protoporphyrin IX, is the ultimate precursor during the biosynthesis of heme. The use of this class of PS was begun at the Mayo Clinic by Lipson et al. (1961), who observed that crude preparations of hematoporphyrin resulted in a red fluorescence of neoplastic tissues observed during surgery. In the attempt to improve the tumor localization, derivatives were obtained by treating blood hematoporphyrin with acetic acid and sulfuric acid, followed by neutralization. This resulted in the formation of dimers and oligomers. The partial purification of these aggregates lead to an enrichment of the active fraction, Photofrin® (Dougherty and Marcus, 1992). Photofrin® was the first PS to be approved for use on humans, and it has been successfully employed in the pretreatment of many types of tumor, including lung, esophageal, cervical, bladder and gastric tumor (Dougherty et al., 1998; Ackroyd et al., 2001; MacDonald and Dougherty, 2001; Dougherty, 2002; Pervaiz and Olivo, 2006). Photofrin® is currently the most important commercial form of hematoporphyrin derivatives. There are several drawbacks of the first-generation porphyrin-based PS: the poorly defined chemical composition; a limited application potential due to the low molar extinction at 630nm, i.e. the absorption wavelength applicable for irradiation in biological tissues; the accumulation not only in tumors, but also, though to a lesser extent, in some normal tissues, including the skin, and the slow clearance from the body, which can last even weeks (Dougherty, 2002).

To enhance the potential of PDT and its clinical applications, a second generation of PS was produced, with properties comparable with (or superior to) Photofrin®, including chemical purity, increased photon absorption at long wavelengths allowing the treatment of deep lying tissue, improved retention in tumor tissues, rapid clearance from surrounding normal tissues, high quantum yield of reactive oxygen species (ROS), and minimal toxicity in the dark (Gomer, 1991; Jori, 1996). The second generation PS, many of which are currently under clinical trials, include naturally occurring and chemically modified or synthesized compounds such as benzoporphyrin derivatives, chlorophyll-a and bacteriochlorophyll-a derivatives, chlorins, phthalocyanines, hypericin and hypocrellins (Sharman et al., 1999; MacDonald and Dougherty, 2001; Pandey et al., 2006). To the second generation PS belongs also protoporphyrin IX (PpIX), produced endogenously upon topical or systemic administration of the natural porphyrin precursor δ-aminolevulinic acid (ALA). A selective localization of PpIX can be obtained depending on morphological, environmental, enzymatic and cell-specific factors (Kennedy et al., 1996; Collaud et al., 2004; Fotinos et al., 2006).

The third generation PS are usually derived from the second generation PS; their specific targeting is obtained by introducing chemical groups to enhance both biological specificity and selectivity (MacDonald and Dougherty, 2001).

Light Sources for Photodynamic Therapy

Visible and near-infrared regions of radiation light are used in PDT to activate phototoxic processes. The choice of the light wavelength required for irradiation is dependent on the absorption properties of the PS used (Dougherty et al., 1998). Especially for treatments *in vivo*, it has to be taken into account that the light penetration in the tissues increases with the wavelength length and depends on the optical properties of the tissues; in addition, the photodynamic effect will depend not only on the light dose and on the PS absorption properties, but also, among others, on the PS intra-tissue localization and oxygen availability.

The choice of the PS and of the light wavelength, in turn, will depend also on the kind of tissue and lesions to be treated: long wavelengths are more suitable to treat deep layers of a tissue, whereas relatively short wavelengths should be preferred when undesired damage effects on an underlying tissue layer are to be avoided (this is the case, for example, with bladder tumours, where damage of the muscular wall would result in a functional impairment of the organ).

Different light sources can be used in PDT. Lasers ensure a powerful monochromatic emission and coherence. These properties can be important in particular for *in vivo* application, since they allow precise focusing and optimal coupling with optical fibres, with a facilitated delivery of optimal doses of light to the treatment sites, particularly important to cure interstitial lesions. Unfortunately, lasers suffer of some drawbacks, such as a limited choice of wavelength for each kind of laser, the high costs and difficult maintenance, justified by sophisticated equipments which are often large and require complex operations. As a consequence, they can often be considered as an over-engineered solution at least for some biological/clinical tasks.

Light sources other than lasers are incandescent lamps, and high-pressure or low pressure arc lamps, all of which covering the entire visible spectrum. Incandescent lamps are essentially conventional light bulbs, and arc lamps contain a gas that conducts electricity at high temperatures: high-pressure arc lamps contain mercury or xenon, while low-pressure arc lamps contain fluorescent material. Common disadvantages of all these sources are the difficulty to focus the light on the target site of treatment, the wide spectrum requiring a selection of the desired wavelength (with a consequent loss of power), and the necessity of filtering to avoid thermal effects. The consequences can be long exposure times, reduced treatment efficacy, and increased risk of undesirable side-effects

Light-emitting diodes (LEDs), have been more recently considered as light sources for PDT (Soldani et al., 2004, Bottone et al., 2006; Soldani et al., 2007a,b). LEDs are semiconductor devices that emit incoherent narrow-spectrum light when electrically biased in the forward direction, giving rise to a form of electroluminescence. LEDs have a relatively narrow bandwidth, usually 20 to 50 nm. The color of the emitted light depends on the chemical composition of the semiconducting material used, and can cover the range of wavelengths from ultraviolet to the visible and the near infrared and infrared spectral region. The flexibility provided by chip-on-board techniques makes it possible to design customized LED illuminators for various PDT applications, with attention to the development of LEDs with increasing power outputs. Compared to lasers and other non-laser sources, LEDs are less expensive, easy to use, limited in size and stable in their emission over a very long time span.

Rose Bengal Acetate is an Efficient Photosensitizer

Rose Bengal (RB) belongs to the series of the xantene dyes, whose photophysical properties are deeply influenced by the aromatic substituents. With respect to the well-known, unsubstituted fluorescein, the presence of chlorines and iodines on the RB xanthene ring induces a marked heavy-atom effect that provides the dye with a high efficiency of intersystem crossing resulting in a high probability of photochemical reactions with high quantum yield of singlet oxygen generation (Neckers and Valdes Aguilera, 1993).

RB at non toxic concentrations is prevented to enter the cells in the absence of a carrier, because of its anionic nature (Croce et al., 1992; Feenstra and Tseng, 1992). The addition of an acetate group to the xanthene rings affects the RB π electron system modifying the photophysical and physico-chemical properties of the fluorophore (Bottiroli et al., 1997): this determines a quenching of both the fluorescence emission and the photosensitizing properties of the native molecule, and enhances the molecule hydrophobicity favoring its intracellular accumulation. In the cells, RB acts as a fluorogenic substrate: once inside the cell, the bound acetate groups are removed by esterases thus restoring the native RB molecule with both the fluorescence and photosensitizing original properties.

On these bases, RB-Ac has been proposed as a promising compound for a new strategy of targeting in PDT (Bottiroli et al., 1997). Incubation of cells with the fluorogenic substrate gives rise to an intracellular accumulation of the active RB molecule, that results from the balance of three defined processes: the substrate influx from the medium into the cells; the enzymatic hydrolysis, which specifically removes the quencher group restoring the original properties of RB; and the product efflux from the cells. These three steps can strongly influence both the extent of RB accumulation and its redistribution during the incubation time. The strict connection of all these aspects with the biological properties of the cells made RB-Ac a promising molecule for the investigation of the relationship between the induction of photosensitization damages at different intracellular sites and the activation of cell death.

In C6 rat glioma cells, RB-Ac administration resulted in a very rapid appearance of intracellular emission by the RB restored molecules, with a cytoplasmic, perinuclear, polar localization that persisted for incubation times longer than 30 min (Bottiroli et al., 1997). In resistant cells expressing the MDR-1 phenotype the increased ability to transport RB outward was counterbalanced by the enhancement of the cell membrane traffic activity, promoting RB-Ac internalization, and by the increase in the esterase activity (Croce et al., 2002). The remarkable phototoxic response induced on cells even under mild treatment conditions and after short incubation time with RB-Ac suggested a rapid localization of the active PS in very sensitive subcellular sites (Bottiroli et al., 1997).

The aim of this article is to review original results by our research group on the effects of photosensitization with RB-Ac on cell organelle structure and function; based on the evidence of multiple organelle photodamage, a mechanism is proposed for the induction of apoptosis by different interacting molecular pathways.

MATERIALS AND METHODS

Cell Model and Experimental Conditions

Human HeLa cells (ATCC, Rockville, USA) were grown in 75 cm^2 flasks in Dulbecco's minimal essential medium supplemented with 10% fetal bovine serum, 1% glutamine, 100 U of penicillin and streptomycin (Celbio s.r.l., Milan, Italy), in 5% CO_2 humidified atmosphere. Twenty four hours before the experiments, cells were seeded on coverslips, in six-multiwell plates for light microscopy, or in 25 cm^2 Petri dishes, for electron microscopy and cytometry.

Cells were incubated with 1 x 10^{-5} M RB-Ac for different times ranging from 1 min to 120 min. Based on these loading experiments, cells were incubated for 60 min before being irradiated at 530 ± 15 nm by means of a light emitting diode (LED; Fraen, Milan, Italy): the total light dose was 1.6 J/cm^2 with an uniform light fluency of 20mW/ cm^2 at the cell surface. After irradiation, cells were transferred to a drug-free complete medium, and allowed to grow for different post-irradiation (p-i) time lengths, from 24 hr to 72 hr.

Morphological and Cytochemical Detection of Organelle Photodamage

The dynamic redistribution of photoactive RB molecules in the cytoplasm of HeLa cells was monitored by confocal fluorescence microscopy.

To detect structural damage of different organelles as a consequence of cell photosensitization with RB-Ac, phase contrast microscopy and transmission electron microscopy were used. Conventional and confocal fluorescence microscopy and flow cytometry were used after fluorochrome-based cytochemical labeling of organelles (endoplasmic reticulum, Golgi Apparatus, mitochondria, cytoskeletal proteins) or of specific cytosolic protein. Some details of the techniques used are given in the figure legends or can be found in the original articles quoted in the text.

RESULTS

Rose Bengal Acetate Redistributes Dynamically Inside the Cells

PS uptake by tumor cells is crucial for an effective PDT. ROS have a short half-life and act on close proximity of their site of generation; therefore, the subcellular localization of the

PS is of special importance, since it determines the occurrence and the extent of the primary damage (Plaetzer et al., 2003).

Using confocal laser fluorescence microscopy, Soldani et al. (2007a, in preparation) demonstrated that, following incubation with RB-Ac, the RB restored molecules redistribute dynamically in the cytoplasm of HeLa cells where they first appear as discrete bright spots (already after 1 min incubation) and then accumulate in a perinuclear-polar pattern of fluorescence (Figure 2). These findings indicate that RB is restored first in the endosomes, in agreement with the RB-Ac hydrophobic properties which promote the adsorption of the substrate to the cell membrane and its internalization *via* the membrane trafficking.

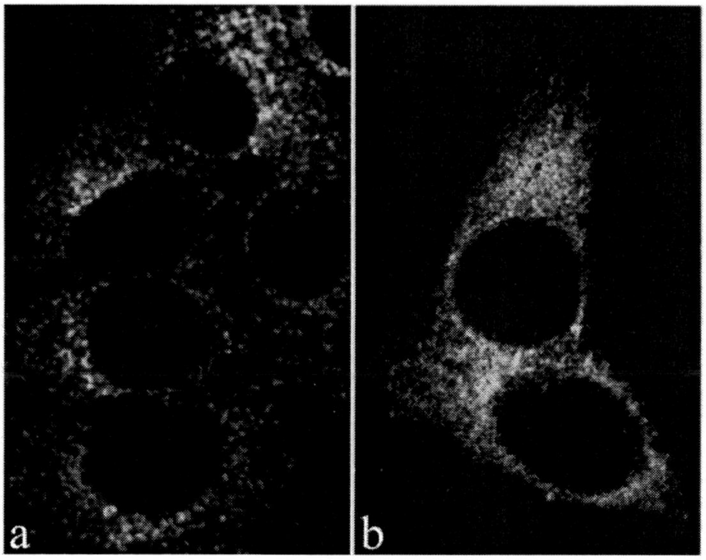

Figure 2. Confocal images of distribution of rose Bengal acetate in HeLa cells: after 5 min (a) fluorescence exhibits a granular, polar-shaped pattern around the nucleus. From 30 min onwards (b) fluorescence is more diffuse in the cytoplasm (for details see Soldani et al., 2007a).

The early RB accumulation pattern is also due to the organelle-related esterase activity, which is known to be especially present in the endoplasmic compartment (Stroemhaug et al., 1998): this was already found to account for the preferential accumulation of RB in multidrug-resistant cells lines from mouse melanoma and human ovarian cancer, compared to the corresponding wild type cell lines (Croce et al., 2002).

RB localization became more diffuse throughout the cytoplasm at 30 min, suggesting that the photoactive molecules can possibly interact with cytosolic structures too. This diffuse fluorescence pattern persisted, without appreciable changes, for the longer incubation times considered in the study (i.e., 120 min): irradiation with a light dose of 1.6 J/cm^2 was therefore performed on cells exposed to RB-Ac for 60 min, when the presence of the fluorescence diffuse pattern indicated the achievement of a dynamic equilibrium in the intracellular redistribution of RB as the result of a balance in the processes RB-Ac accumulation, enzyme hydrolysis and product diffusion/efflux (Bottiroli et al., 1997).

Several Different Organelles Undergo Structural and Functional Damage in Cells Photosensitized with RB-Ac

Light and electron microscopy showed that, following photosensitization of HeLa cells with RB-Ac, the endoplasmic reticulum (ER) and the Golgi apparatus (GA) were significantly affected (Soldani et al., 2004, 2007a).

After staining with ER-Tracker (a photostable fluorochrome that is selective for the ER in living cells: Figure 3), apparent changes were observed in the organization of the vacuolar system, especially at 24 hr p-i when the ER was prominently enlarged and swollen compared to the controls; this was confirmed at transmission electron microscopy. At 48-72 hr p-i, the cells became shrunken with more densely packed cisternae; moreover, in photodamaged cells, the cytoplasm appeared more electron-dense than in controls with clusters of free ribosomes (Soldani et al., 2007a). These modifications were observed also by immunofluorescence analysis, using an antibody that recognizes the glucose regulated protein 94 (GRP 94), a resident protein of ER (unpublished observation).

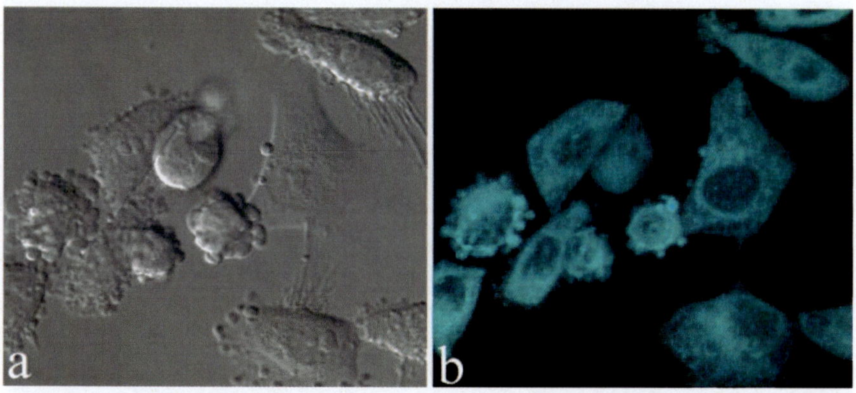

Figure 3. RB-Ac-treated HeLa cells 72h p-i. In living cells, phase contrast (a) shows alterations in the cell shape, with emission of filopodia and massive surface blebbing. The same cells, stained with ER-Tracker and observed at confocal microscopy (b; for details see Soldani et al., 2007a).

After immunolabeling with an autoimmune serum recognizing the Golgi-specific protein, golgin-97 (Figure 4), the typical ribbon-like distribution pattern of cisternae was observed around the nucleus in control cells; in RB-Ac-treated cells 24hr p-i, the cisternae of the GA became packed, and after 48-72 hr p-i they were more fragmented and scattered throughout the cytoplasm. The change in the organization of the Golgi cisternae was confirmed at electron microscopy, where they appeared more densely stacked than in untreated cells at 24hr p-i interval, to become progressively more dilated with large scattered vesicles at longer p-i times (Soldani et al., 2004).

The intracellular redistribution of RB in a perinuclear and polar-shaped pattern suggests that PS molecules do follow the route of endosomes, which are known to undergo maturation and acidification in the perinuclear region (Lakadamyali et al., 2003). The retrograde transport of endocytosed exogenous compounds from the trans-Golgi network to the ER (Sandvig et al., 1992; Mallet and Maxfield, 1999) could explain the relocation of RB within the whole vacuolar system.

Both microtubules and actin cytoskeleton have been reported to be possible targets of photosensitization processes (Berg and Moan, 1997). In particular, Conlon et al. (2002) suggested that actin microfilaments might undergo light-induced disruption in cells photosensitized by RB. Consistently, we found in RB-Ac treated and irradiated HeLa cells that actin microfilaments formed bundles which became progressively thicker, especially at the cell periphery, for increasing p-i times (Soldani et al., 2007a). In parallel with this intracellular restructuring of microfilaments, alterations of the overall cell shape were apparent in living cells already after relatively short p-i times (30 min): most of the cells emitted filopodia and the cytoplasm appeared extensively vacuolated; for longer p-i times, the morphological changes became more prominent, and the cell surface was covered by long microvilli and cytoplasmic fragments were released at the cell surface (Figure 3). This was shown by phase-contrast time-lapse microscopy (unpublished observation) and by scanning electron microscopy (Soldani et al., 2006). Also the microtubules reorganized into thick bundles, especially at 24-48 hr p-i (see Figure 4): the restructuring of this cytoskeleton components can account, at least in part, for the dramatic changes in the organization of the ER and GA described above.

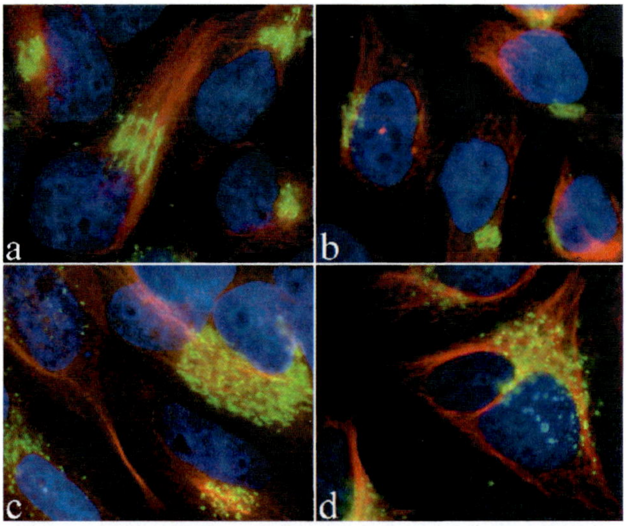

Figure 4. Digital micrographs of control (a) and RB-Ac-treated (b-d) HeLa cells after dual immunolabeling of the Golgi apparatus (autoimmune serum, green fluorescence) and α-tubulin (red); DNA staining with Hoechst 33258 (blue). The cisternae of the Golgi apparatus appear initially packed (b: 24hr p.i.), and then fragment and scatter throughout the cytoplasm (c,d: 48, 72hr p.i; for details see Soldani et al., 2004).

After incubation with RB-Ac and irradiation, also the mitochondria showed morphological and functional modifications. In untreated HeLa cells, the mitochondria are uniformly distributed throughout the cytoplasm (Figure 5); at 24h p-i, they clustered around the nucleus; at 48h p-i they were more condensed and larger than in controls; and at 72h p-i, mitochondria formed dense masses in the cytoplasm. The analysis at electron microscopy showed that, at 24h p-i, mitochondria were smaller and more electron-dense than in untreated cells, with densely packed cristae, whereas at 48h p-i they became swollen and assumed a balloon-like and vacuolated morphology; sometimes, the outer membrane broke down, leaving a single layer of membrane with disorganized cristae. At 72h p-i interval, dense and irregularly shaped precipitates were observed, which were often filling the whole inner space (Bottone et al., 2006).

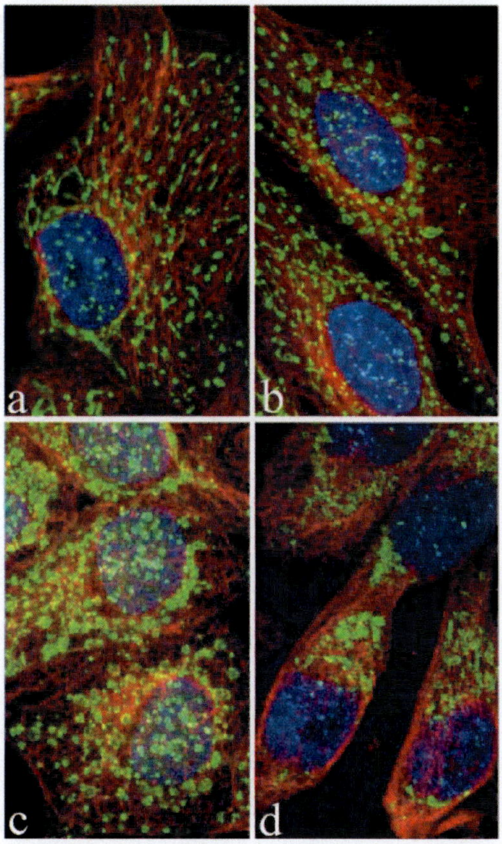

Figure 5. Confocal images of control (a) and RB-Ac-treated (b-d) HeLa cells after dual immunolabeling of mitochondria (autoimmune serum, green fluorescence) and α-tubulin (red); DNA staining with Hoechst 33258 (blue). In control (a), the mitochondria were uniformly distributed throughout the cytoplasm; at 24h p-i (b) and more at 48h p-i (c) they are larger than in controls and more clustered around the nucleus. At 72h p-i, mitochondria form dense masses in the cytoplasm (d; for details see Bottone et al., 2006).

The flow cytometric analysis after vital staining with Mitotracker green (a cell-permeant vital dye that accumulates in active mitochondria) demonstrated that, after photosensitization of HeLa cells with RB-Ac and irradiation, the active mitochondrial mass progressive decreased for increasing p-i intervals (Bottone et al., 2006).

The quantitative and morphological alterations observed after drug exposure and irradiation were paralleled by apparent changes in the mitochondrial potential (ΔΨm): this was measured using the fluorescent probe, JC-1 which exhibits a potential-dependent accumulation in mitochondria (Molecular Probes, Invitrogen). Bottone et al. (2006a) demonstrated that in untreated HeLa cells a single cell population is present, in which the JC-1 red-to-green ratio is rather homogeneous; at 24h p-i interval, and even more apparently at

48h and 72h, a collapse of the mitochondrial potential was observed, as indicated by the shift to higher green fluorescence signals of a large fraction of cells. Interestingly, observation at fluorescence microscope showed that the distribution of red and green mitochondria was not homogeneous inside the cytoplasm, depolarized green mitochondria being especially clustered around the nucleus and the red ones at the cell periphery (Bottone et al., 2006).

Induction of Apoptotic Cell Death

In parallel with the different organelle alterations described above, several cells exhibited morphological signs of apoptosis, with nuclear shrinking, chromatin condensation and karyorrhexis; this was especially observed at 72h p-i interval (Figure 6).

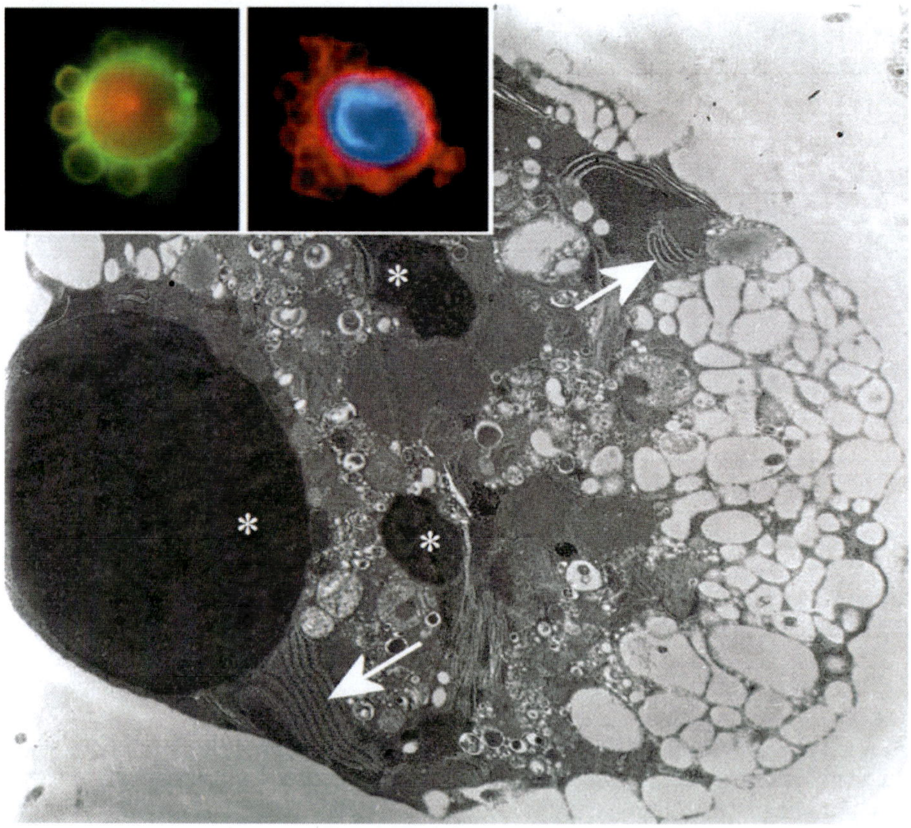

Figure 6. Electron micrograph of a late apoptotic RB-Ac-treated HeLa cell 72h p-i. Dilated vescicles and densely packed cysternae (arrows) are present in the vacuolated cytoplasm; the nuclear fragments (asterisks) contain highly condensed chromatin. Inset: late apopotic HeLa cells after dual-staining with FITC-conjugated annexin V and propidium iodide (left) or after immunolabeling of actin (re., 2007b).d fluorescence) and DNA counterstaining with Hoechst 33258 (for details, see Soldani et al

It is known (Fadok et al., 1992) that already in early apoptosis phosphatidylserine residues are translocated from the inner phospholipids leaflet to the outer leaflet of the plasma membrane, and can bound Annexin-V under vital conditions; in later apoptotic stages, the function of the plasma membrane is compromised and the dyes, such as propidium iodide, which are exclude under vital conditions, can enter the cell (Figure 6, inset). Consistently, flow cytometric measurements after dual-staining with FITC-conjugated Annexin V and propidium iodide demonstrated that the percentage of cells exhibiting apoptotic features increased with increasing p-i times (Soldani et al., 2007b).

Caspases are a group of cysteine proteases that exist within the cell as inactive pro-forms or zymogens and are generally considered as main effectors of apoptosis. Following the induction of apoptosis, these zymogens can be cleaved to form active enzymes which degrade both structural and regulatory proteins in the nucleus and cytoplasm (Earnshaw, 1995; Faleiro and Lazebnik, 2000).

We observed by immunofluorescence analyses that the active forms of caspase 12, caspase 9 and caspase-3 can be detected after RB-Ac photosensitization, especially in the samples after 72h p-i (Figure 7). Caspase-3 is a key player in the execution phase of apoptosis, and is responsible for the cleavage of a large number of proteins (Porter and Janicke, 1999) including the DNA fragmentation factor (DFF) and poly(ADP-ribose) polymerase 1 (PARP-1). In our experimental system the increase of caspase-3 positive cells was strictly correlated to an increase of cells positive to p89 (Figure 7), a protein fragment generated by the caspase-specific cleavage of PARP-1 (reviewed in Soldani and Scovassi, 2002). PARP-1 cleavage is responsible for the inactivation of the poly(ADP-ribosylation) process and may precede the typical endonucleolytic cleavage of nuclear DNA (this was found in etoposide-treated human HeLa cells: Soldani et al., 2001).

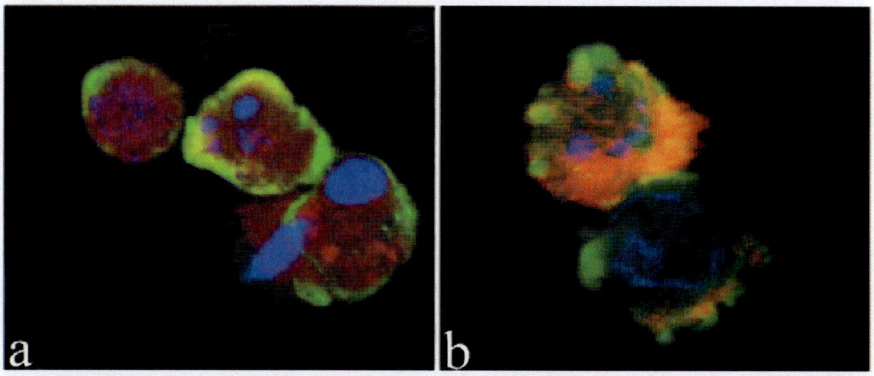

Figure 7. Confocal micrographs of RB-Ac-treated HeLa cells 72h p-i: green fluorescence immunolabeling is for active caspase 3 (a) or for the p89 fragment of PARP-1 (b). The cytoplasm was labeled with Evans Blue and the DNA counterstained with Hoechst 33258.

Fluorescence DNA labeling and electron microscopy showed that chromatin undergoes the typical apoptotic condensation in HeLa cells after photosensitization with RB-Ac (Figure 7); however, we found that no intranucleosomal cleavage of DNA takes place, as revealed by the TUNEL assay and agarose gel electrophoresis (Soldani et al., 2007b). On the other hand, the flow cytometric analysis of DNA content showed the presence of a sub-G1 peak in DNA histograms at 72h p-i interval: this suggests that, under the treatment conditions used, a solubilization of DNA has taken place, irrespective of the absence of low molecular weight fragments. The fragmentation of DNA into nucleosomal lengths has been widely recognized as a molecular hallmark of apoptosis (Bursch et al., 1990); however, high molecular weight DNA fragmentation seems to be a more reproducible apoptotic phenomenon (Sleiman et al., 1998) and several studies have demonstrated its role as an essential initial step in apoptosis (Brown et al., 1993; Oberhammer et al., 1993; Walker et al., 1994).

CONCLUSION

It is well known that apoptosis may be induced by photodamage upon administration of PS and irradiation (Oleinick et al., 2002; Ali and Olivo, 2003; Ma et al., 2003). Apoptosis is strictly related to the physiological "programmed cell death", which plays key roles during the embryonic development as well as in the maintenance of tissue homeostasis, at the end of histogenesis (Ranganath and Nagashree, 2001; Opferman and Korsmeyer, 2003). In apoptosis, cells die in a controlled, regulated fashion in response to a variety of stimuli and are rapidly removed by phagocytes without major perturbation of the tissue integrity: because of this, the induction of apoptosis has gained attention (especially in recent years) for its potential application in therapy (Guchelaar et al., 1997; Schmitt and Lowe, 1999; Morgan and Oseroff, 2001; Lebedeva et al., 2003; Adachi et al., 2004).

The induction of apoptosis by PDT depends on several factors: the PS concentration and the light dose applied represent important external factors, while the cellular metabolic state as well as the cell cycle phase can be named among major internal parameters influencing the sensitivity of cells to photodynamic treatment (Wyld et al., 2001). In fact, cell death results from the photodamage of target organelles and the intracellular location of the PS molecules is essential for the onset of the apoptotic process.

In the case of RB, the intracellular accumulation of photoactive molecules upon RB-Ac administration results from the balance of three defined processes, i.e. the substrate influx, enzymatic hydrolysis, and product efflux from the cells (Bottiroli et al., 1997; Croce et al., 2002). Under our experimental conditions, RB molecules are rapidly restored in the cytoplasm where they initially locate in peripheral vesicles, then redistribute in a perinuclear, polar-shaped pattern to finally diffuse in the cytosol. This intracellular redistribution of RB suggests that the PS molecules first transit through the route of endosomes, which are known to undergo processes of maturation and acidification in the perinuclear region (Lakadamyali et al., 2003). The retrograde transport of endocytosed compounds from the endosomes to the ER *via* the trans Golgi network (Sandvig et al., 1992; Mallet and Maxfield, 1999) can account for the relocation of RB within the whole vacuolar system; herein, remarkable alteration were observed in the organization of both the GA and ER, at different p-i times. Actually, the sensitivity of these organelles to the photodamage has also been reported for other PS (e.g., hypocrellin B and its derivatives, hypericin, toluidine blue, temoporfin: Miller et al., 1997; Uzdensky et al., 2001; Siboni et al., 2002).

RB molecules were also observed to diffuse in the cytosol; irradiation can therefore determine the production of ROS in close proximity of cytosolic structures or the membranes of organelles other than the ER and GA.

In fact, after RB-Ac administration and irradiation, microtubules and microfilaments too were found to undergo rearrangements, with the formation of thick bundles especially at the cell periphery, contemporary with the occurrence of changes at the cells surface, such as the formation of microvilli followed by extensive blebbing and release of cytoplasmic fragments. According to the literature, actin cytoskeleton seems to play multiple roles in relation with the apoptotic phenomena. Its engagement as a primary effector of apoptosis through activation of a caspase-3-dependent mechanism has been proposed based on experiments of selective cytoskeleton disruption (Suria et al., 1999). However, the generally prevailing concepts concern an involvement of actin cytoskeleton in phenomena which are typical of the progression and execution of apoptosis, such as blebbing, and nuclear fragmentation, with an influence on the recognition and removal of apoptotic bodies through phagocytosis (May and Machesky, 2001). In addition, actin and microtubules might perform redundant or cooperative roles in apoptotic fragmentation, depending on their relative amounts in different cell types (Moss and Lane, 2006).

Flow cytometric analyses demonstrated that photosensitization of HeLa cells with RB-Ac results in the collapse of mitochondrial potential, as revealed by JC-1 fluorochromization, and leads to a decrease in the active mitochondrial mass, as determined by Mitotracker green staining (Bottone et al., 2006); all these changes have been observed to occur in apoptotic cells (Reipert et al., 1995; Lee et al., 2005). Fluorescence and electron microscopy also revealed ballooning of the mitochondria and disruption of the internal mitochondrial structure; similar results were obtained by Luchetti et al. (2006) on apoptotic U937 promonocytic leukemia cells exposed to UV-B and by Huang et al. (2005) on HL60 cells; similarly swollen mitochondria have been described by Shen et al. (2002) in apoptotic esophageal carcinoma cells treated with arsenic trioxide, and these morphological changes of mitochondria always preceded the characteristic nuclear modifications found in late apoptosis.

The experimental results reviewed in the present article demonstrate that, unlike other PS which have been described as organelle-specific, RB proved to be able to damage different targets, likely due to its capability to relocate dynamically to different intracellular sites, when administered as the fluorogenic substrate, RB-Ac.

The first targets of RB photosensitization have been found to be the ER and GA. The proper function of the ER is essential for cell survival, and any perturbation of its structure/function induces cellular damage and may result in apoptosis (Oyadomari et al., 2002). Various conditions can disturb ER functions and such events are collectively termed *ER stress*. This stress includes inhibition of protein glycosylation, reduction of formation of disulfide bonds, calcium depletion from the ER lumen, impairment of protein transport from the ER to the GA, expression of misfolded proteins, and so on (Oyadomari et al., 2002; Dahmer, 2005). With its central location and pivotal role in membrane traffic, the GA is in an ideal position to sense and integrate information about the state of the cell (Rios and Bornens, 2003), and the continuous trafficking between the ER and GA may allow integrated apoptotic signalling between these compartments during cell death induction through photodamage; consistently, stresses applied onto these structures can trigger the apoptotic cascade, *via* the activation of either regulatory proteins or directly effective apoptotic proteases, such as caspase 12 or caspase 2 (Nakagawa et al., 2000; Ferri and Kroemer, 2001; Boyce and Yuan, 2006; Kessel et al., 2006). Activated caspase-12 has been reported to specifically cleave caspase-9, which then cleaves caspase-3 (Rao et al. 2002): this is a key player in the execution phase of apoptosis, and is responsible for the cleavage of a large number of proteins (Porter and Janicke, 1999) including the DNA fragmentation factor (DFF) and poly(ADP-ribose) polymerase 1 (PARP-1), an enzyme that participates in DNA repair by transferring ADP-ribose polymers to histones and other nuclear proteins (Stroh and Schultze-Osthoff, 1998). In RB-Ac-treated and irradiated HeLa cells, the presence of caspase-3 positive cells was strictly related to an increase of cells immunopositive for p89, a protein fragment generated by caspase-specific cleavage of PARP-1. Cleavage of PARP-1 is responsible for the inactivation of the poly(ADP-ribosylation) process (Scovassi et al., 1998; Scovassi and Poirer, 1999; Soldani and Scovassi, 2002) and thereby further promotes the demise of the cell.

The PDT-induced cleavage of Golgi proteins can promote apoptotic signalling, determines the blockade of membrane traffic (Sesso et al, 1999; Lane et al., 2002), and certainly contributes to the organelle disassembly in apoptotic cells which is required for packaging the GA remnants into apoptotic blebs (Machamer, 2003). Chiu et al (2002) also demonstrated that caspase-mediated proteolysis of key vesicle tethering factors (p115) contributes to GA breakdown during apoptosis and may act to propagate downstream apoptotic signals.

GA fragmentation also occurs in viable cells when microtubules are depolymerised, or when the microtubule motor protein dynein is inhibited (Lane and Allan, 1998). A similar scenario is found in the apoptotic cytoplasm: dynein is inhibited by caspase cleavage of its intermediate chains and the p150glued component of the accessory complex dynactin (Lane et al., 2001). Although microtubules coordinate the dynamics of GA in healthy cells (Lane and Allan, 1998), it remains unexplored whether their disassembly (or dynein inhibition) contributes to these changes in organelle properties. Certainly, photodamage of microtubules as it occurs in RB-Ac treated and irradiated cells can synergistically contribute to the observed restructuring of the GA.

The actin cytoskeleton and microtubules can also undergo complex interactions with mitochondria during apoptosis induction, as discussed by Suria et al. (1999).

Mitochondria have been identified as the central processing organelles in most apoptotic pathways. Various apoptotic signals from either outside or inside the cell converge in mitochondria which can sense stress signals and initiate (when appropriate) the execution phase of apoptosis (Fleury et al., 2002). The mitochondrial stress pathway (even in the case of damage induced by mitochondria-specific PS) begins with the loss of the transmembrane potential and the release from mitochondria to the cytoplasm of pro-apoptotic polypeptide factors (among which cytocrome c; the apoptosis-inducing factor, AIF; and smac/DIABLO) which contribute to the formation of the apoptosome protein complex (Verhagen et al., 2000). In particular, cytochrome c interacts with Apaf-1, causing self-cleavage and activation of caspase-9 (Fan et al., 2005) which further cleaves downstream caspases, including caspase-3, and leads to the loss of the plasma membrane integrity and finally to chromatin condensation and karyorrhexis (Green, 2000; Kandasamy et al., 2003).

In the case of RB-induced apoptosis, we can hypothesize that its induction can possibly follow different, though interacting routes, one apoptotic signal starting from the ER and Golgi, and another involving the mitochondrial pathway. The resulting dramatic disruption of the whole cytoplasmic organization would drive cells to death and prevent completion of

DNA degradation into oligonucleosome fragments. The fragmentation of DNA into nucleosomal lengths has been widely recognized as a molecular hallmark of apoptosis (Bursch et al., 1990); however, DNA degradation into high molecular weight fragments seems to be a more reproducible apoptotic phenomenon (Sleiman et al., 1998) and several studies have demonstrated its role as an essential step in apoptosis (Brown et al., 1993; Oberhammer et al., 1993; Walker et al., 1994).

The success of PDT, as a therapeutic approach, actually depends on its selectivity against tumor cells and on the number of cells which can be damaged to death. Fluorogenic substrates, such as RB-Ac, derived by chemical modification of PS proved to be very promising agents for PDT, as they accumulate in larger amount in tumor cells than the native PS molecule, and the restored photoactive form persists inside the cells for longer times (Bottiroli et al., 1997): this is a great advantage for the application *in vivo*, since lower drug doses can be used, thus reducing the risk of acute cytotoxicity in the absence of irradiation and the negative effects of systemic toxicity.

Our results on a cell model system in culture also demonstrated that administration of RB-Ac results in the ability of photoactive RB to damage different cell organelles/structures, when a dynamic equilibrium in its intracellular redistribution is achieved: this allows apoptosis to be induced through the cooperative action of different triggering stimuli, thus increasing the probability that cells can be driven to controlled death. This mode of cell death is especially convenient for improving the efficacy of PDT *in vivo*: in fact, the preferential induction of apoptosis entails the rapid and safe clearance of apoptotic cells by phagocytes, whereas the massive presence of necrotic cells is undesired, because of their less effective removal and their capability to induce an inflammatory response.

Acknowledgements

The authors wish to thank Mr.s Paola Veneroni for the technical assistance in cell culture and Dr. Claudia Alpini (IRCCS San Matteo, Pavia, Italy), for the kind gift of the autoimmune sera recognizing a Golgi protein and mitochondria.

Confocal fluorescence images were taken at the Centro Grandi Strumenti of the University of Pavia. This work was supported by the University of Pavia (FAR 2005) and MIUR (PRIN 2005058254).

References

Adachi S, Leoni LM, Carson DA, Nakahata T. Apoptosis induced by molecular targeting therapy in hematological malignancies. *Acta Haematol.* 2004; 111: 107-123.

Ackroyd R, Kelty C, Brown N, Reed M. The history of photodetection and photodynamic therapy. *Photochem. Photobiol.* 2001; 74:656-669.

Ali SM, Olivo M. Mechanisms of action of phenanthroperylenequinones in photodynamic therapy. *Int. J. Oncol.* 2003; 22:1181-1191.

Berg K, Moan J. Lysosomes and microtubules as targets for photochemotherapy of cancer. *Photochem. Photobiol.* 1997; 65:403-409.

Bonnett R. Photodynamic therapy in historical perspective. *Rev. Contemp. Pharmacother.* 1999; 10:1-17.

Bottiroli G, Croce AC, Balzarini P, Locatelli D, Baglioni P, Lo Nostro P, Monici M, Pratesi R. Enzyme-assisted cell photosensitization: A proposal for an efficient approach to tumor therapy and diagnosis. The Rose Bengal fluorogenic substrate. *Photochem. Photobiol.* 1997; 66:374-383.

Bottone MG, Soldani C, Fraschini A, Alpini C, Croce AC, Bottiroli G, Pellicciari C. Enzyme-assisted photosensitization with Rose Bengal acetate induces structural and functional alteration of mitochondria in HeLa cells. *Histochem.Cell Biol.* 2006; in press PMID 17024456 a.

Boyce M, Yuan J. Cellular response to endoplasmic reticulum stress: a matter of life or death. *Cell Death Differ.* 2006; 13:363-373.

Brown DG, Sun XM, Cohen GM. Dexamethasone-induced apoptosis involves cleavage of DNA to large fragments prior to internucleosomal fragmentation. *J. Biol. Chem.* 1993; 268:3037-3039.

Brown SB, Mellish KJ. Verteporfin: a milestone in opthalmology and photodynamic therapy. *Expert Opin. Pharmacother.* 2001; 2:351-361.

Bursch W, Kleine L, Tenniswood M. The biochemistry of cell death by apoptosis. *Biochem. Cell Biol.* 1990; 68:1071-1074.

Chiu R, Novikov L, Mukherjee S, Shields D. A caspase cleavage fragment of p115 induces fragmentation of the Golgi apparatus and apoptosis. *J. Cell Biol.* 2002; 159:637-648.

Collaud S, Juzeniene A, Moan J, Lange N. On the selectivity of 5-aminolevulinc acid-induced protoporphyrin IX formation. *Curr. Med. Chem. Anticancer Agents* 2004; 4:301-316.

Conlon KA, Rosenquist T, Berrios M. Site-directed photochemical disruption of the actin cytoskeleton by actin-binding Rose Bengal-conjugates. *J. Photochem. Photobiol.* 2002; 68: 140-146.

Croce AC, Wyroba E, Bottiroli G. Distribution and retention of rose bengal and disulphonated aluminium phthalocyanine: a comparative study in unicellular eukaryote. *J. Photochem. Photobiol.* 1992; 16:319-330.

Croce AC, Supino R, Lanza KS, Locatelli D, Baglioni P, Bottiroli G. Photosensitizer accumulation in spontaneous multidrug resistant cells: a comparative study with Rhodamine 123, Rose Bengal acetate and Photofrin. *Photochem. Photobiol. Sci.* 2002; 1:71-78.

Dahmer MK. Caspases-2, -3, and -7 are involved in thapsigargin-induced apoptosis of SH-SY5Y neuroblastoma cells. *J. Neurosci. Res.* 2005; 80:576-583.

Dolmans DE, Fukumura D, Jain RK. Photodynamic therapy for cancer. *Nat. Rev. Cancer* 2003; 3:380–387.

Dougherty TJ, Marcus SL. Photodynamic therapy. *Eur. J. Cancer* 1992; 10:1734-1742.

Dougherty TJ, Gomer CJ, Henderson BW, Jori G, Kessel D, Korbelik M, Moan J, Peng Q. Photodynamic therapy. *J. Natl. Cancer Inst.* 1998; 90:889-905.

Dougherty TJ. An update on photodynamic therapy applications. *J. Clin. Laser Med. Surg.* 2002; 20:3-7.

Earnshaw WC. Nuclear changes in apoptosis. *Curr. Opin. Cell Biol.* 1995; 7:337-343.

Fadok VA, Voelker DR, Campbell PA, Cohen JJ, Bratton DL, Henson PM. Exposure of phosphatidyl serine on the surface of apoptotic lymphocytes triggers specific recognition and removal by macrophages. *J. Immunol.* 1992; 148:2207–2216.

Faleiro L, Lazebnik Y. Caspases disrupt the nuclear-cytoplasmic barrier. *J. Cell Biol.* 2000; 151:951-959.

Fan TJ, Han LH, Cong RS, Liang J. Caspase family proteases and apoptosis. *Acta Biochim Biophys. Sin* 2005; 37:719-727.

Feenstra RPG, Tseng SCG. What is actually stained by rose bengal? *Arch. Ophthalmol.* 1992; 110:984-993.

Ferri KF, Kroemer G. Organelle-specific initiation of cell death pathways. *Nat. Cell Biol.* 2001; 3: 255-263.

Fotinos N, Campos MA, Popowycz F, Gurny R, Lange N. 5-aminolevulinc acid derivatives in photomedicine: Basics, applications and perspectives. *Photochem. Photobiol.*, in press, PMID: 16542112.

Fleury C, Mignotte B, Vayssiere JL. Mitochondrial reactive oxygen species in cell death signaling. *Biochimie* 2002; 84:131-141.

Gomer CJ. Preclinical examination of first and second generation photosensitizers used in photodynamic therapy. *Photochem. Photobiol.* 1991; 54:1093-1107.

Green DR. Apoptotic pathways: paper wraps stone blunts scissors. *Cell* 2000; 102:1-4.

Guchelaar HJ, Vermes A, Vermes I, Haanen C. Apoptosis: molecular mechanisms and implications for cancer chemotherapy. *Pharm. World Sci.* 1997; 19:119-125.

Hopper C. Photodynamic therapy: A clinical reality in the treatment of cancer. *Lancet Oncol.* 2000; 1:212-219.

Huang HF, Chen YZ, Wu Y. Mitochondria-dependent apoptosis induced by a novel amphipathic photochemotherapeutic agent ZnPcS2P2 in HL60 cells. *Acta Pharmacol. Sin.* 2005; 26:1138–1144.

Ibbotson SH. Topical 5-aminolaevulinic acid photodynamic therapy for the treatment of skin conditions other than non-melanoma skin cancer. *Br. J. Dermatol.* 2002; 146:178-188.

Jori G. Tumour photosensitizers: approaches to enhance the selectivity and efficiency of photodynamic therapy. *J. Photochem. Photobiol. B* 1996; 36:87-93.

Kandasamy K, Srinivasula SM, Alnemri ES, Thompson CB, Korsmeyer SJ, Bryant JL, Srivastava RK. Involvement of proapoptotic molecules Bax and Bak in tumor necrosis factor-related apoptosis-inducing ligand (TRAIL)-induced mitochondrial disruption and apoptosis: differential regulation of cytochrome c and Smac/DIABLO release. *Cancer Res.* 2003; 63:1712-1721.

Kennedy JC, Marcus SL, Pottier RH. Photodynamic therapy (PDT) and photodiagnosis (PD) using endogenous photosensitization induced by 5-aminolevulinic acid (ALA); mechanisms and clinical results. *J. Clin. Laser Med. Surg.* 1996; 14:289-304.

Kessel D, Vicente MGH, Reiners JJ Jr. Initiation of apoptosis and autophagy by photodynamic therapy. *Lasers Surg. Med.* 2006; 38:482-488.

Lakadamyali M, Rust MJ, Babcock HP, Zhuang X. Visualizing infection of individual influenza viruses. *Proc. Natl. Acad. Sci. USA* 2003; 100:9280-9285.

Lane J, Allan V. Microtubule-based membrane movement. *Biochim. Biophys. Acta* 1998; 29:1376:27-55.

Lane JD, Vergnolle MA, Woodman PG, Allan VJ. Apoptotic cleavage of cytoplasmic dynein intermediate chain and p150(Glued) stops dynein-dependent membrane motility. *J. Cell Biol.* 2001; 153:1415-1426.

Lane JD, Lucocq J, Pryde J, Barr FA,Woodman PG, Allan VJ, Lowe M. Caspase-mediated cleavage of the stacking protein GRASP65 is required for Golgi fragmentation during apoptosis. *J. Cell Biol.* 2002; 156:495-509.

Lebedeva IV, Su ZZ, Sarkar D, Fisher PB. Restoring apoptosis as a strategy for cancer gene therapy: focus on p53 and mda-7. Semin. *Cancer Biol.* 2003; 13:169-178.

Lee CF, Liu CY, Hsieh RH, Wei YH. Oxidative stress-induced depolymerization of microtubules and alteration of mitochondrial mass in human cells. *Ann. N.Y. Acad. Sci.* 2005; 1042:246-254.

Lipson R, Baldes E, Olsen A. The use of a derivative hematoporphyrin in tumour detection. *J. Natl. Cancer Inst.* 1961; 26:1-8.

Luchetti F, Canonico B, Curci R, Battistelli M, Mannello F, Papa S, Tarzia G, Falcieri E. Melatonin prevents apoptosis induced by UV-B treatment in U937 cell line. *J. Pineal Res.* 2006; 40:158-166.

Ma L, Tai H, Li C, Zhang Y, Wang Z-H, Ji W-Z. Photodynamic inhibitory effects of three perylenequinones on human colorectal carcinoma cell line and primate embryonic stem cell line. *World J. Gastroenterol.* 2003; 9: 485-490.

MacDonald IJ, Dougherty TJ. Basic principles of photodynamic therapy. *J. Porhyrins and Phthalocyanines* 2001; 5:105-129.

Machamer CE. Golgi disassembly in apoptosis: cause or effect? *Trends Cell Biol.* 2003; 13:279-281.

May RC, Machesky LM. Phagocytosis and the actin cytoskeleton. *J. Cell Sci.* 2001; 114:1061-1077.

Mallet WG, Maxfield FR. Chimeric forms of furin and TGN38 are transported from the plasma membrane to the Trans-Golgi Network via distinct endosomal pathways. *J. Cell Biol.* 1999; 146: 345-359.

Miller GG, Brown K, Ballangrud AM, Barajas O, Xiao Z, Tulip J, Lown JW, Leithoff JM, Allalunis-Turner MJ, Mehta RD, Moore RB. Preclinical assessment of hypocrellin B and hypocrellin B derivatives as sensitizers of photodynamic therapy of cancer: progress update. *Photochem. Photobiol.* 1997; 65:714-722.

Moan J, Berg K. The photodegradation of porphyrins in cells can be used to estimate the lifetime of singlet oxygen. *Photochem. Photobiol.* 1991; 53:449-553.

Morgan J, Oseroff AR. Mitochondria-based photodynamic anti-cancer therapy. *Adv. Drug Deliv. Rev.* 2001; 49:71-86.

Moss DK, Lane JD. Microtubules: forgotten players in the apoptotic execution phase. *Trends Cell Biol.* 2006; 16:330-338.

Nakagawa T, Zhu H, Morishima N, Li E, Xu J, Yankner BA, Yuan J. Caspase-12 mediates endoplasmic-reticulum-specific apoptosis and cytotoxicity by amyloid-beta. *Nature* 2000; 403:98-103.

Neckers DC, Valdes-Aguilera OM. Photochemistry of the xanthene dyes. *Adv. Photochem.* 1993; 18:315-394.

Nestor MS, Gold MH, Kauvar AN, Taub AF, Geronemus RG, Ritvo EC, Goldman MP, Gilbert DJ, Richey DF, Alster TS, Anderson RR, Bank DE, Carruthers A, Carruthers J, Goldberg DJ, Hanke CW, Lowe NJ, Pariser DM, Rigel DS, Robins P, Spencer JM, Zelickson BD. The use of photodynamic therapy in dermatology: results of a consensus conference. *J. Drugs Dermatol.* 2006; 5:140-154.

Oberhammer F, Wilson JW, Dive C, Morris ID, Hickman JA, Wakeling AE, Walker PR, Sikorska M. Apoptotic death in epithelial cells: cleavage of DNA to 300 and/or 50 kb fragments prior to or in the absence of internucleosomal fragmentation. *EMBO J.* 1993; 12:3679-3684.

Oleinick NL, Morris RL, Belichenko I. The role of apoptosis in response to photodynamic therapy: what, where, why, and how. *Photochem. Photobiol. Sci.* 2002; 1:1-21.

Opferman JT, Korsmeyer SJ. Apoptosis in the development and maintenance of the immune system. *Nat. Immunol.* 2003; 4:410-415.

Oyadomari S, Araki E, Mori M. Endoplasmic reticulum stress-mediated apoptosis in pancreatic beta-cells. *Apoptosis* 2002; 7:335-345.

Pandey RK, Goswami LN, Chen Y, Gryshuk A, Oseroff JR, Dougherty TJ. Nature: a rich source for developing multifunctional agents. Tumour-imaging and photodynamic therapy. *Lasers Surg. Med.* 2006; 38:445-467.

Peng Q. Editorial: photodynamic therapy and detection. *J. Environ. Pathol. Toxicol. Oncol.* 2006; 25:1-6.

Pervaiz S, Olivo M. Art and science in Photodynamic therapy. *Clin. Exp. Pharmacol. Physiol.* 2006; 33:551-556.

Plaetzer K, Kiesslich T, Krammer B, Hammerl P. Characterization of the cell death modes and the associated changes in cellular energy supply in response to AlPcS4-PDT. *Photochem. Photobiol. Sci.* 2002; 1:172-177.

Porter AG, Janicke RU. Emerging roles of caspase-3 in apoptosis. *Cell Death Differ.* 1999; 6:99-104.

Ranganath RM, Nagashree NR. Role of programmed cell death in development. *Int. Rev. Cytol.* 2001; 202:159-242.

Rao RV, Castro-Obregon S, Frankowski H, Schuler M, Stoka V, del Rio G, Bredesen DE, Ellerby HM. Coupling endoplasmic reticulum stress to the cell death program. An Apaf-1-independent intrinsic pathway. *J. Biol. Chem.* 2002; 277:21836-21842.

Reipert S, Berry J, Hughes MF, Hickman JA, Allen TD. Changes of mitochondrial mass in the hemopoietic stem cell line FDCP-mix after treatment with etoposide: a correlative study by multiparameter flow cytometry and confocal and electron microscopy. *Exp Cell Res.* 1995; 221:281-288.

Rios RM, Bornens M. The Golgi apparatus at the cell centre. *Curr. Opin. Cell Biol.* 2003; 15:60-66.

Sandvig K, Garred O, Prydz J, Kozlov JV, Hansen SH, van Deurs B. Retrograde transport of endocytosed Shiga toxin to the endoplasmic reticulum. *Nature* 1992; 358:510-512.

Sesso A, Fujiwara DT, Jaeger M, Jaeger R, Li TC, Monteiro MM, Correa H, Ferreira MA, Schumacher RI, Belisario J, Kachar B, Chen EJ. Structural elements common to mitosis and apoptosis. *Tissue Cell* 1999; 31:357-371.

Schmitt CA, Lowe SW. Apoptosis and therapy. *J. Pathol.* 1999; 187:127-137.

Scovassi AI, Denegri M, Donzelli M, Rossi L, Bernardi R, Mandarino A, Frouin I, Negri C. Poly(ADP-ribose) synthesis in cells undergoing apoptosis: an attempt to face death before PARP degradation. *Eur. J. Histochem.* 1998; 4:251-258.

Scovassi AI, Poirier GG. Poly(ADP-ribosylation) and apoptosis. *Mol. Cell. Biochem.* 1999; 199:125-137.

Sharman WM, Allen CM, van Lier JE. Photodynamic therapeutics: basic principles and clinical applications. *Drug Discovery Today* 1999; 4: 507-517.

Shen ZY, Shen WY, Chen MH, Shen J, Cai WJ, Zeng Y. Mitochondria, calcium and nitric oxide in the apoptotic pathway of esophageal carcinoma cells induced by As2O3. *Int. J. Mol. Med.* 2002; 9:385-390.

Siboni G, Weitman H, Freeman D, Mazur Y, Malik Z, Ehrenberg B. The correlation between hydrophilicity of hypericins and helianthrone: internalization mechanisms, subcellular distribution and photodynamic action in colon carcinoma cells. *Photochem. Photobiol. Sci.* 2002; 1:483-491.

Sleiman RJ, Catchpoole DR, Stewart BW. Drug-induced death of leukaemic cells after G2/M arrest: higher order DNA fragmentation as an indicator of mechanism. *Br. J. Cancer* 1998; 77:40-50.

Soldani C, Bottone MG, Pellicciari C, Scovassi AI. Two-color fluorescence detection of Poly (ADP-Ribose) Polymerase-1 (PARP-1) cleavage and DNA strand breaks in etoposide-induced apoptotic cells. *Eur. J. Histochem.* 2001; 45:389-392.

Soldani C, Scovassi AI. Poly(ADP-ribose) polymerase-1 cleavage during apoptosis: an update. *Apoptosis* 2002; 7:321-328.

Soldani C, Bottone MG, Croce AC, Fraschini A, Bottiroli G, Pellicciari C. The Golgi apparatus is a primary site of intracellular damage after photosensitization with Rose Bengal acetate. *Eur. J. Histochem.* 2004; 48:443-448.

Soldani C, Croce AC, Bottone MG, Fraschini A, Biggiogera M, Dini L, Pellicciari C, Bottiroli G. Rose Bengal Acetate as a fluorogenic substrate for photosensitization of cultured tumor cells: subcellular location, organelle photodamage and induction of apoptosis. In preparation, 2007a

Soldani C, Bottone MG, Fraschini A, Croce AC, Bottiroli G, Scovassi AI, Pellicciari C. Photosensitization by rose Bengal acetate induces apoptosis without oligonucleosomic DNA fragmention. In preparation, 2007b.

Stroemhaug PE, Berg TO, Fengsrud M, Seglen PO. Purification and characterization of autophagosomes from rat hepatocytes. *Biochem. J.* 1998; 335:217-224.

Stroh C, Schulze-Osthoff K. Death by a thousand cuts: an ever increasing list of caspase substrates. *Cell Death Differ.* 1998; 5:997-1000.

Suria H, Chau LA, Negrou E, Kelvin DJ, Madrenas J. Cytoskeletal disruption induces T cell apoptosis by a caspase-3- mediated mechanism. *Life Sci.* 1999; 65:2697-2707.

Uzdensky AB, Ma LW, Iani V, Hjortland GO, Steen HB, Moan J. Intracellular localisation of hypericin in human glioblastoma and carcinoma cell lines. *Lasers Med. Sci.* 2001; 16;276-283.

Verhagen AM, Ekert PG, Pakusch M, Silke J, Connolly LM, Reid GE, Moritz RL, Simpson RJ, Vaux DL. Identification of DIABLO, a mammalian protein that promotes apoptosis by binding to and antagonizing IAP proteins. *Cell* 2000; 102:43-53.

Walker PR, Weaver VM, Lach B, Le Blanc J, Sikorska M. Endonuclease activities associated with high molecular weight and internucleosomal DNA fragmentation in apoptosis. *Exp. Cell Res.* 1994; 213:100-106.

Wyld L, Reed MW, Brown NJ. Differential cell death response to photodynamic therapy is dependent on dose and cell type. *Br. J. Cancer* 2001; 84:1384–1386.

In: Cell Apoptosis and Cancer
Editor: Albina W. Taylor, pp. 47-60

ISBN: 978-1-60021-506-3

Chapter III

BIOLOGY OF APOPTOSIS AS PREDETERMINED CARCINOGENESIS

Lawrence M. Agius*

Department of Pathology, St Luke's Hospital, Gwardamangia, University of Malta Medical School, MSIDA, Malta, Europe

ABSTRACT

In terms of an evolutionary biology of apoptosis pathways as cascade events it is significant that a full range of operative factors would implicate the dynamics of phenotypic variability characteristic of infiltrating neoplasia. It is towards the definition of such phenotypic variability that subsequent events prove or otherwise refute the concept of genetic instability as an operative framework in the development and potentiation of specific carcinogenetic pathways as predetermined biology of malignant lesions. It is particularly in view of a whole series of events borne out by the infiltration of tissues and spread of tumor cells as clumps of such cells that carcinogenesis would be purely definable as a predetermination of apoptotic versus anti-apoptotic events. These would operate in both initiation and subsequent characterization of lesion neogenesis as malignant disease. Only in such terms can there evolve a full series of events in the further definition of pathways as characteristic of both initiation and promotion of carcinogenetic agent action in further evolution of apoptosis versus anti-apoptosis transformation in the biology of neoplastic lesions.

* Correspondence concerning this article should be addressed to Lawrence M. Agius, Department of Pathology, St Luke's Hospital, Gwardamangia, University of Malta Medical School, MSIDA, Malta Europe. 27 Ballarat Guzeppi Caruana Str., Tal-Virtu, Rabat, RBT09 Malta Europe. Tel.: 356-21451752, (Work): 356-2595-1663; Fax: 356-21224286; Email: lawrence.agius.um.edu.mt, lawrence.agius@gov.mt.

Introduction: Postnatal Consequence of Transformed Prenatal Apoptosis of Ovarian Germ Cells

Prenatal apoptosis of ovarian germ cells would not be simply transferred as an extension to the postnatal period but there would be realized an apparent stress-related series of events that provoke transformed induction of postnatal events [1].

One might view dynamics of evolved change in apoptosis of germ cells as an illustration of how stress events are integral consequences of a promotion of persistent loss of germ cells during postnatal life. In a sense, understanding how pathologic insults are themselves involved in such stress induction would account for how postnatal apoptosis is transformed consequence of prenatal apoptosis of the ovarian germ cells.

Endothelial Differentiation Receptivity

Endothelial differentiation-related receptivity appears an integral mechanism central to the evolving effects of lysophosphatidic acid (LPA) or of lysolipid phosphate that affects proliferation of cells as a form of survival of these cells [2]. Cell proliferation of cells would be considered as a basic entity in terms of cell survival and might implicate endothelial cell biology that undergoes a series of transformations based on viability issues.

Cells may proliferate in order to survive and might involve a series of direct consequences arising as endothelial cell-induced events and viability mechanisms. The consequences of cell proliferation would implicate dynamics of cell survival as a direct consequence of such cell proliferation [30].

Formation of Matrix Metalloproteinases

Upregulation of the concentration of MMP-9 (matrix metalloproteinase-9) as via cytokines and growth factors would titer out the action of Tissue Inhibitors of Metalloproteinases (TIMP) in a manner linked fundamentally to cascade proteolysis [32] as induced by plasmin [3]. The role of pro-MMP-9 bound to TIMP would also appear to effectively invoke proteolysis especially in the activation of matrix metalloproteinases as a synthetic pathway in the first instance.

One might speak of the formation of matrix metalloproteinases and of their subsequent action on matrix in an essential overall functionality involving proteolysis linked directly to ingress of metastatic clumps of tumor cells into blood vessel lumina. One would have to consider how cascade proteolysis is an integral function of the matrix metalloproteinases in tumor cell spread and metastasis in terms of a direct function of their formation through proteolysis. Resistance of FasL-triggered apoptosis may be triggered by MMP-7 in colorectal carcinoma [19].

Mediated P53 Mutation Effect

Specific subcellular localization of p53 appears to develop often in the absence of p53 mutation [4]. This would account for arrest in the G1 phase of the cell cycle and for apoptosis of the cell in a manner related to p53 mutation. It might be significant to consider how p53 mutation is a mediated form of subcellular localization of the p53. N and S forms of neuroblastoma would be distinguished largely in terms of the strict sublocalization of p53 to the nucleus in the N form in a manner that would link its accumulation with a failure of ubiquitin-mediated proteolysis of the p53.

The observed failed arrest at G1 in some cases of neuroblastoma would perhaps signify ineffective sublocalization of p53 arising in terms of nuclear membrane fragmentation as due to irradiation effect [41].

PTEN in Dysplastic Transformation

The protean manifestations of PTEN as a tumor suppressor gene would appear linked to early malignant transformation as the dysplastic stage of endometrial adenocarcinoma and to a plethora of effects determining metastatic spread as a late event [5]. PTEN and p53 have many different regulatory functions including apoptosis [31]. One might speak of epigenetic effects that predetermine dysplastic development in a context of consequential progression as acquired metastatic potential manifested only in later stages of tumor growth and spread.

Cowden's disease as a constitutional deletion/mutation of the PTEN is an approximation of sporadic cases of deletion/mutation of PTEN that involve multiple organs and also hamarthomas of brain and skin. In this sense, hamartomas are simply transcriptional failure in tissue differentiation central also to neoplastic transformation.

S100 Protein-Negative Schwann Cell Derivation

An intricate histogenic heterogeneity of neurofibromas appears intrinsically related to Schwann cell origin with deletions of neurofibromin gene [6].

One might speak of how plexiform neurofibromas arise as Schwann cell proliferations that subsequently acquire increasing malignant potentiality consistent with loss of the S100 protein reactivity of the neoplastic cells. S100 nonreactivity of Schwann cells would indicate a progression of proliferative pools of cells that not only become undifferentiated but also coincidentally acquire malignant attributes of proliferative and infiltrative spread.

It is perhaps in this sense of a coincident proliferative and infiltrative spread that plexiform neurofibroma becomes an S100 protein –negative population of Schwann cell –derived neoplastic cells.

Specific Blood Vessel Invasion Mechanics

Serum levels of matrix metalloproteinases would appear linked to production by macrophages or tissue histiocytes and perhaps also by circulating blood monocytes that subsequently infiltrate foci of tissue injury or areas of tissue repair or foci of infiltrating neoplasia [7]. Matrix metalloproteinases such as gelatinase B or A would be produced by colorectal carcinoma in a manner integral to an overall phenomenon of invasion of local tissue and of blood vessels in metastatic spread. Gelatinase would act specifically in blood vessel spread of metastatic tumor cell clumps to allow for a better delineation not only of prognostic indices in tumor growth and invasion but particularly for better visualization of metastatic spread in blood vessel breakdown at an enzymatic tissue level.

Antiapoptosis Linked to Angiogenesis

Apoptosis and tumor necrosis would specifically correlate with failed angiogenesis and especially with suppressed proliferation of tumor cells [8]. Inhibitors of matrix metalloproteinases would operate to best suppress both angiogenesis and tumor cell proliferation as two primary modes of action in suppressing metastatic spread and infiltration by tumor cells and also in inducing apoptosis and necrosis as two derived modes of action.

Tumor cell spread develops concurrently with angiogenic responses that define matrix proteolysis as a series of events linked to tumor cell proliferation and antiapoptosis. Indeed, antiapoptosis might be linked inherently to angiogenesis in tumor cell development and subsequent spread [46]. Synergistic activity of many genes appears implicated in metastatic tumor cell selectivity [18]

Pro-Apoptosis and Apoptosis as Biologic Progressiveness

Expression of increased apoptosis would indicate an index activity concurrently progressive with an increasingly aggressive biology of meningiomas that are specifically metastatic and infiltrative [9]. Anaplastic and atypical meningiomas are variably graded aggressive biologic forms of a neoplastic type that is usually low-grade infiltrative and specifically non-metastatic. One would recognize how increased apoptosis is not simply a receptor-ligand binding phenomenon but a responsive pathway system whereby morphologic anaplasia or atypia converge as pathways of increasingly aggressive biologic activity, as induced also by FAT10. FAT10 is a ubiquitin-modifier protein implicated in mitosis and apoptosis, cytokine production and antigen presentation [33].

Meningiomas with increased mitotic activity are a system of progressive anaplasia that is specifically proapoptotic in a manner conducive to biologic aggressiveness, anaplasia and atypia. Active participation of infiltrative behavior of tumor cells towards metastatic potential might indicate an active role for proapoptosis and active apoptosis leading to neoplastic aggressiveness.

This would be progressively reflected in increasing anaplasia of the meningioma cells.

Fas receptor or ligand binding would implicate increasing mitotic activity and the directly operative antiapoptosis in terms of progression of a meningioma that is aggressive biologically as early recurrence and spread of tumor cells.

TRANSNUCLEAR MEMBRANE TRANSLOCATION

Intranuclear localization of p73 with intracytoplasmic localization of its antagonist p53 [29] would correspond to an undifferentiated neuroblastic profile [10]. P53 as an aberrant intracytoplasmic accumulation would correspond to loss of suppressor gene function in a manner that would relate to accumulation intraneuronally of a stable deltaN-p53 form alien to transient transcriptional activity potential.

Nuclear-cytoplasmic translocation of p73 with corresponding intranuclear accumulation of p53 would prove a correlate of neuroblastic differentiation in a manner that accounts for better prognostic outcome in patients with more differentiated neuroblastomas as affecting the adrenal medulla.

The strict role of p73 in terms of correlative antagonism to the suppressor gene action of p53 might correlate especially with the active creation of aberrant forms of the p53 that accumulate intracytoplasmically with oncogenesis and with decreasing differentiation of the neuroblastoma. It is with reference to the transcription –activation domain of p73 that there would develop a negative impact on p53 transactivation of target genes.

Intranuclear accumulation of the N-truncated p73 might correlate with the stabilization of an isoform of the p73 that would correspondingly combine in a stable manner with p53 leading to loss of the gene suppressor function.

Intranuclear accumulation of p73 and hence p73 exclusion from the cytoplasm would correlate with lost suppressor gene function of p53 and would relate to an essential translocation abnormality between nucleus and cytoplasm that is specifically oncogenic [17].

Such a defect in transnuclear membrane translocation would prove oncogenic and be suggestive of mechanisms related to the emergence of a stable isoform as represented by the delta N-p73 dominant form. This would induce sustained transcriptional effects that are specifically not transitory.

The creation of multiple splice forms of p73 distinct from the normal outcome of p53 processing would be suggestive of a multi-combinatory series of steps whereby the creation of stabilized delta-N-p73 is safeguarded against according to cellular and nuclear homeostatic mechanisms.

The active creation of multiple isoforms of p73 would constitute mechanisms that prevent the outcome of a stable delta-N-p73 thus effectively promoting oncogenesis as seen with emergence of undifferentiated neuroblastic tumor. p53 is aberrant and intracytoplasmic in location in terms of a significantly failed transnuclear translocation of such aberrant p53. This would be primarily an active factor towards stabilization effect of p73 in the creation of a sustained transcriptional effect. Such sustained transcription would specifically correlate with a neuroblastic oncogenesis that implicates undifferentiated neuroblastic biologic potential. P53 functions may be altered and mediated by cutaneous human papillomavirus type 38 and DeltaNp73 in human carcinogenesis [21].

PHENOTYPIC VARIABILITY IN TRANSFORMATION

Protein S100 A4 appears to integrally incorporate invasiveness with a tumor progressiveness that permissively allows and also actively promotes mechanistic pathways of transformation as typical of the acquisition of the malignant phenotype [11]. Once acquired, such malignant phenotype appears characterized by a propensity for undergoing further phenotypic transformation, and numerous genotypic modality effects are presumably involved and underlie such phenotypic variability in progression and transformation.

Combined malignant phenotypic progression and transformation are integrated within a unified framework of neoplastic cell proliferation, invasion and metastatic spread. Apoptosis would contribute to genetic stability by removing cells with damaged DNA [42]. A model of overall carcinogenesis would universally account for dedifferentiation and biologic aggressiveness as labile neoplastic attributes. Increasing stage and grade of a neoplasm such as astrocytoma would involve basic model systems as typified by S100 proteins, especially S100A4. Conceptual considerations of cell cycle entry and progression of tumor cells would correlate with a proliferative framework linked to cell adhesion interactivity and modulated cytoskeletal dynamics that are capable of progression.

One strict pathobiologic attribute of neoplastic tissue is its particular propensity to self-progress as a lesion that grows in an infiltrative and metastasizing pattern of replacement of cells and tissues not otherwise capable of significant phenotypic transformation, as through the action of estrogen receptor beta in ovarian carcinogenesis [44].

INTERACTIONS OF EXTRACELLULAR MATRIX AND GLIOMA CELLS

Infiltrative gliomas would primarily constitute non-metastasizing neoplasms largely in terms of how interactive adhesion phenomena develop with regard to the extracellular matrix and any aggregate gliomatous cell interactions [12]. Angiogenesis as a prominent feature of most forms of infiltrative glioma would constitute a substitute phenomenon that supplements other supporting systems in widely infiltrating high-grade gliomas. Active and inactive cell-cell interactions and apoptosis of the tumor cells alternating with essential antiapoptosis in glioma cell progression and genesis would appear implicated. Upregulation of matrix metalloproteinase 7 is associated with increased resistance to FasL-tirggered apoptosis [37].

The interactivities between glioma cells with fibronectin and vimentin rather than with collagen might constitute a mechanism whereby infiltrativeness progresses further as a metastasizing form of proliferative tumor cell growth.

Proliferation of tumor cells might actually constitute a mode of interactivity with the extracellular matrix that is modulated specifically towards a propensity for the onset and progression of infiltrative growth [15,16,45].

A self-progressiveness in infiltrative growth would implicate a proliferation of glioma cells that is inherently spreading in scope and actual development. Infiltrative glioma growth is simply active or inactive in terms of migratory behavior of individual cells and of increased apoptotic rate of tumor cells inherently implicating antiapoptosis in the genesis of gliomas.

Apoptosis as a Summation Phenomenon

Apoptosis refers to a summation phenomenon borne out by conflicting processes involved in the possible progression of a given neoplastic lesion [14]. One such process would adopt the targeted individual cell involvement as a form of progressive pathology that is both transforming and incremental in nature. It is in view of such contrasting events that paradoxically induce transformation and increments in progression of a neoplastic process that there would develop evolutionary pathways related closely to abnormal development of antiapoptotic pathways [22,40].

The ING1 gene is involved in regulation of apoptosis and the cell cycle and acts as a candidate suppressor gene; toegether with ING2, its functions depend on activity of p53 [20].

Involvement of such an integrative transformation pathway might perhaps guarantee the evolution of a neoplastic event that is singularly progressive and yet also one primarily designed to maintain a state of persistent genetic instability. NR4A1, NR4A2 and NR4A3 are steroid nuclear hormone receptors involving immediate early genes implicated in cell mitogenic responses and apoptosis [34]. Integral incorporation of genomic events that are mainly amplifying in nature might help characterize a neogenesis as a singly integrating process of transforming identity of pathways affecting self-proliferation and spread in tissues beyond just biologic interactivity with stroma or other neoplastic cells.

Vascular blood supply as an original source for such integrative transformation in neoplastic genesis would be consistent with a singular process of promoted participation in subsequent development of a unique carcinogenetic event.

Increased Gene Transcription as Antiapoptosis

A central role for increased nuclear levels of beta-catenin in colonic carcinogenesis would appear a particularly effective approach towards understanding how loss of intercellular adhesion relates with infiltrative behavior and with antiapoptosis and proliferative activity [13].

Such a basic scheme of induced loss of intercellular adhesion at the tight junction level might implicate nuclear transcription as an increased pathway of promoter induction that would not implicate amplification or mutation of genes of transcription such as TCF-4 or AXIN. Antiapoptosis would constitute a characterized pathway for such increased gene transcription arising also in the pathogenesis of carcinogenesis.

Loss of intercellular adhesion would account for metastatic potentiality arising primarily as a characterized determinant of antiapoptosis and internuclear promoter induction in increased gene transcription.

Anti-Apoptosis and Mitotic Activity

Apoptotic index in breast carcinoma would somehow constitute a correlate with other prognostic factors in patients with the neoplasm that however is not of independent prognostic status [23].

One might view various pathologic parameters of breast carcinoma such as high histologic grade, presence of metastases, high mitotic rate and even absent estrogen-progesterone receptor status as simply aspects of a pathobiology of transformation that do not arise directly from simple cellular dynamics of the malignant transformation event itself.

Malignant transformation would constitute a level of operability that is well beyond simple consideration of just cellular features or of cellular parameters of progression in breast carcinoma. Several genes are synergistically active for the selection process in several metastatic lesions [35].

Apoptosis in breast carcinomatous lesions might in various ways be related to how infiltrative events for example are related to a high mitotic rate of proliferative activity. Proteolytic activity and matrix remodeling are important in tumor progression [43].

A high mitotic rate might clearly establish indices of anti-apoptotic systems of progression as in terms of histological grade and infiltrativeness, that would redefine inter-relationships of cell division with prognosis and with effective carcinogenesis.

Considerations of apoptosis versus anti-apoptosis as an integral process of progression in carcinogenesis might help clarify how patient prognosis is a complex interplay of mitotic activity and of anti-apoptosis rather than of just an apoptotic index parameter.

P53 Amplification in In Situ Breast Carcinoma

Overamplification of p53 appears intrinsically linked to an increased incidence of p53 mutations [24]. Indeed, one might consider how increased copy production of p53 would result in mutated p53.

In terms of how in situ carcinoma of the breast evolves to a combined lesion of in situ and invasive component carcinoma there might arise a system of essential evolution of p53 mutation that is overexpressed in the invasive component, based on a clonal system of progression.

It is perhaps in terms of a clonal amplification that one might conceive of in situ carcinoma that invades in parallel with increasing amplification of p53 and of p53 mutations.

Increased p53 production with accumulation in the cytoplasm of the tumor cells might perhaps reflect pathways of acquisition of invasive potential that primarily arise from p53 mutation rate.

P53 mutations would arise within in situ lesions and would be clonally amplified in foci of an invasive tumor that is associated with an in situ carcinoma primarily in terms of proliferative events. It is perhaps in terms of clonality of both proliferative and antiapoptotic events that amplifications would prove a variant of mutational progression of the p53 in progressive breast carcinogenesis. Amplified p53 would constitute a variation in the overall increased incidence of p53 mutation rate as arising initially in in situ breast carcinoma and subsequently transformed to an at time accompanying invasive component of the lesion.

Defective DNA Binding Underlies Genetic Instability

Unsilencing of gene expression as through unmethylation or hypomethylation of DNA of heterochromatin might represent effective abnormalities of DNA binding that significantly contribute to a genetic state of instability [25].

Recognition of an unstable genetic state in carcinogenesis might implicate a fundamental defect in DNA binding underlying not simply abnormalities in gene expression but also a strong tendency for constitutive variability in gene expression involving both mutational/deletional/translocation events and amplification of gene copies. It is perhaps in terms of phenomena related to estrogen receptivity of breast carcinomatous cells that one might better understand how amplification events would arise in a context of variable degrees of differentiation of the neoplastic cells in terms of evolving genetic instability.

Genetic instability specifically promoting an amplification of deletions and mutations and translocations of DNA heterochromatin might simply evolve as a basic integral disturbance of gene expression due to impaired DNA binding. Hypomethylation would reflect such a basically central defect in DNA binding in carcinomatous cells that generates genetic instability.

Increased Mitochondrial Membrane Barrier

Pro-apoptosis would paradoxically operate in terms particularly of anti-apoptosis towards transformation involving the bypassing of mechanisms of effective apoptosis [26]. Downregulation affects ubiquitous mitochondrial creatine kinase, promoting loss of apoptosis [36].

A series of mechanisms such as breakdown of the mitochondrial membrane barrier in promoting cell apoptosis through systems such as osmotic disruption might perhaps more significantly be suggestive of whole pathway effects targeting an increased membrane barrier functionality versus dysfunctionality promoting clonogenic survival of tumor cells.

It is perhaps in this sense that clonogenic proliferation of neoplastic cells arises primarily and subsequently progresses as largely increased degrees of mitochondrial membrane barrier promoting effective prevention of release of cytochrome C and also prevention of activation of the caspase cascade events.

Indeed, internal and extrinsic systems of apoptosis might largely depend on a participation of initiator and effector caspase systems of receptor ligand binding that promotes or modulates nuclear transcription.

A basic scheme of intracellular and intercellular communication largely based on a real participating role on the part of the cytoskeleton would perhaps evolve in terms of increased mitochondrial membrane barrier dysfunction in promoting clonogenic proliferation of neoplastic cells. Also, mutation of mitochondrial DNA may serve as an early indication of potential carcinogenesis [39].

Amplified Transcription of Promoter Gene Sequences

A working formulation of genetic factors in carcinogenesis might subserve mechanisms of change based on gene amplification [27].

The various concepts of suppressor genes and of oncogenes would perhaps have to be formulated in terms of a single pathway of inducible transformation of the cell cycle based on attributes of gene amplification. Viruses such as human papilloma virus and its E7 transcription factor might implicate a series of binding systems that operate within binding schemes affecting suppressor genes such as RB1 or neurofibromatosis gene 1 (NF1) and also implicate an antiapoptotic system of progression allied closely to cyclins and cyclin kinases.

Essential coupling of cyclins and cyclin kinases with a series of amplified and mutated genes would involve promote gene sites that involve particularly specific attributes of an increased copy number of promoter regions of genes.

In this way, neoplastic cells are cells with primarily greatly amplified factors via binding/unbinding cycles of suppressor genes in the strict context of blocked apoptosis. Indeed, anti-apoptosis would involve a system of amplified transcription of promoter genes that may be either oncogenes or suppressor genes but whose pro-carcinogenetic effects rely mainly on the stage and restriction points reached by that particular cell towards a mitotic cycle and cell spindle formation sequence.

Phosphorylation and dephosphorylation of suppressor genes such as RB1 and the various pathways of cyclin and cyclin kinase binding steps would determine transcription factor kinetics within a system of availability dependent on cycle stage attributes of the proliferating cell that would or would not transform to a neoplastic cell.

Multigenesis as Genomic Expression Towards Induced Cell Cycling

Mutational change as both a generative pathway of new genetic constitutional expression as well as a system of predisposition towards novel expression would perhaps constitute inherently a non-evolutionary pathway specifically in terms of possibly predetermined biologic endpoint. It is indeed in terms of genetic mutation that develops subsequent to various events of nonrandom type that one might view adaptability of cells as a secondary consequence of nonrandom selectivity reflecting genomic structure and function.

These would operate more than any characteristic attributes of the selectivity process itself as an inducing agent.

One would consider mutation generation a phenomenon arising largely as an attribute of a cellular genome that is essentially active especially in the nondividing cell rather than during active progression of the cell cycle.

One might specifically consider mutagenesis an active characterization of genomic participation especially in the context of nondividing cells in a stationary phase of noncycling. Accumulative damage from mutagenesis would translate in terms of pathways of generative effect towards possible subsequently active cell cycling and active cell dividing activity [38].

Cell Cycling and Antiapoptosis

Aspects of how cell cycle arrest substitute for cell death pathways in terms of essential features of apoptosis would implicate an escape of mechanistic pathways of induced cell death or of progressive tendencies toward cell degeneration and cell death [28]. An overall scheme would incorporate cell cycle activity as a simple integration of multiple processes that actively prevent cell death. The main mechanistic effects of anti-apoptosis would constitute these same mechanistic processes of cell cycling. A system of tumor gene suppression and of suppressor gene function might be involved.

Cell cycling would not only culminate in cell division but more importantly involves active prevention of cell degeneration and cell death. It might be reasonable to identify active anti-apoptosis as a main mechanistic pathway of developmental progressiveness of tumor proliferation as a central originator of neoplastic biology even beyond a strict cellular basis of interpretation and of etiology/pathogenesis.

Conclusion

It appears essentially relevant to consider anti-apoptosis as a fundamental correlate of onset dynamics and subsequent progression of events in carcinogenesis. It would be in terms of a variability of suppression of genes as part of the genetic instability of neoplasia that various pro-apoptotic agonists are variably suppressed within a context particularly of an increased mitochondrial barrier. The evolution of proliferative activity of tumor cells as a manifestation of a centrally operative antiapoptotic influence would prove an important function of cell cycling activity that actively prevents cell degeneration and cell death. Anti-apoptosis would appear a central attribute that mechanistically promotes the evolution of pathways of carcinogenesis in the final emergence of the variable neoplastic phenotype of individually characterized tumors such as in situ and invasive breast carcinoma.

REFERENCES

[1] Moreta Y, Tilly JL "Sphingolipid regulation of female gonadal cell apoptosis" *Annals NY Acad Sci* 2000;905(1):209-220.

[2] Goetzl EJ, Lee H, Dolezalova H, Kalli KR et al. "Mechanisms of lysolipid phosphate effects on cellular survival and proliferation" *Annals NY Acad Sci* 2000;905(1):177-187.

[3] Hahu-Dantona E, Ramos-DeSimone H, Sipley J, Nagase H et al. "Activation of Pro MMP-9 by a Plasmin/MMP-3 cascade in a tumor cell model: regulation by tissue inhibitors of metalloproteinases" *Annals NY Acad Sci* 1999;678(1):372-387.

[4] Tweddle DA, Malcolm AJ, Cole M, Pearson ADJ, Lunec J "P53 cellular localization and function in neuroblastoma" *Am J Pathol* 2001;158(6):2067-2077.

[5] Matter GI "PTEN, a protean tumor suppressor" *Am J Pathol* 2001;158(6):1895-1898.

[6] Perry A, Roth KA, Bauerjee R, Fuller CE, Gutmann DH "NF1 deletions in S100 protein positive and negative cells of sporadic and neurofibromatosis-1 (NF1)-associated plexiform neurofibromas and malignant peripheral nerve sheath tumors" *Am J Pathol* 2001;159(1):57-61.

[7] Zucher S, Hymowitz M, Conner C, Zarrabi HM et al. "Measurement of matrix metalloproteinases and tissue inhibitors of metalloproteinases in blood and tissues: clinical and experimental applications." *Annals NY Acad Sci* 1999;878(1):212-227.

[8] Shalmsky DR, Brekken J, Zou H, McDermott CD et al. "Broad antitumor and antiangiogenic activities of AG3340, a potent and selective MMP inhibitor undergoing advanced oncology clinical trials" *Annals NY Acad Sci* 1999;878(1):236-2270.

[9] Weisberg S, Ashkenazi E, Israel Z, Attia M et al. "Anaplastic and atypical meningiomas express high levels of Fas and undergo apoptosis in response to Fas ligation" *Am J Pathol* 2001;159(4):1193-1197.

[10] Douc-Rasy S, Barrois M, Echeynne M, Kaghad M et al. "delta-p73 alpha accumulates in human neuroblastic tumors" *Am J Pathol* 2002;160(2):631-639.

[11] Mazzucchelli L "Protein S100A4: too long overlooked by pathologists?" *Am J Pathol* 2002;160(1):7-13.

[12] Long AJ, Padarathsingh M, Louis DN, Rempel S, Ladisch S, Gladson C "New approaches to the molecular biology, classification and therapy of nervous system tumors." *Am J Pathol* 2001;159(5):1971-1974.

[13] Wong NAC-S, Pignatelli M "Beta-catenin—A linchpin in colorectal carcinogenesis?" *Am J Pathol* 2002;160(2):389-401.

[14] Gomes CP, Andrade LA "PTEN and p53 expression in primary ovarian carcinomas: immunohistochemical study and discussion of pathogenetic mechanisms" *Int J Gynaecol Cancer* 2006 Jan-Feb;16 Suppl:254-8.

[15] Wu XZ, Chen D, Xie GR "Extracellular matrix remodeling in hepaotcellular carcinoma: effects of soil on seed?" *Med Hypotheses* 2006 Feb24 [Epub ahead of print].

[16] Lak TT, Duran Alonso MB, Van Noorden CJ "Antiprotease therapy in cancer: host or not?" *Expert Opin Biol Ther* 2006 Mar;6(3):257-79.

[17] Zhang DW, Jeang KT, Lee CG "p53 negatively regulates the expression of FAT10, a gene upregulated in various cancers" *Oncogene* 2006 Feb27 [Epub ahead of print].

[18] Mendez O, Martin B, Sanz R, Aragues R, Moreno V, Oliva B eta l. "Under-expression of transcriptional regulators is common in metastatic breast cancer cells overexpressing Bcl-xL" *Carcinogenesis* 2006 Feb 20 [Epub ahead of print].

[19] Wang WS, Chen PM, Wang HS, Liang wY, Su Y "Matrix metalloproteinase-7 increases resistance to Fas-mediated apoptosis and is a poor prognostic factor of patients with colorectal carcinomas" *Carcinogenesis* 2006 Feb10 [Epub ahead of print].

[20] Okano T, Gemma A, Hosoya Y, Hosomi Y, Nara M, Kohubo Y et al. "Alterations in novel candidate tumor suppressor genes, ING1 and ING2 in human lung cancer" *Oncol Rep Mar*;15(3):545-9.

[21] Accardi R, Dong W, Smet A, Cui R, Hautefeuille A, Gabet AS et al. "Skin human papillomavirus type 38 alters p53 functions by accumulation of DeltaNp73" *EMBO Rep* 2006 Jan 6;[Epub ahead of print].

[22] Cheung HH, Arora V, Korneluk RG "Abnormalities of cell structures in tumors: apoptosis in tumors" *EXS* 2006;(96):201-21.

[23] Lipponeu "Apoptosis in breast cancer: relationship with other pathological parameters" *Endocrine and Related cancer* 1999;6:13-16.

[24] Lukas J, Niu N, Pess MF "p53 mutations and expression in breast carcinoma in situ" *Am J Pathol* 2000;156:183-191.

[25] Tsuda H, Takarabe T, Kanai Y, Fukutomi T, Hirohashi S "Correlation of DNA hypomethylation at pericentromeric regions of chromosomes 16 and 1 with histological features and chromosomal abnormalities of human breast carcinomas" *Am J Pathol* 2002;161(3):859-866.

[26] Reed JC "Mechanisms of apoptosis" Warner-Lamber/Parke Davis Award Lecture. *Am J Pathol* 2000 Nov;157(5):1415-1430.

[27] Zwarthoff EC Chapter 2 "Regulatory pathways in normal and tumor cells" Vinkens PJ, Bryn GW (Editors) *Handbook of Clinical Neurology*: Vol 67 Neuro-oncology (Ed.) Vecht CJ.

[28] Gene Cards "p53".

[29] Pluta A, Nyman U, Joseph B, Robak T, Zhivotovsky B, Smolewski P "The role of p73 in hematological malignancies" *Leukemia* 2006 Mar 16 [Epub ahead of print].

[30] Thomadaki H, Scorilas A, Hindmarsh JT "BCL2 family of apoptosis-related genes: functions and clinical implicatons in cancer" *Crit Rev Clin Lab Sci* 2006;43(1):1-67.

[31] Gomes CP, Andrade LA "PTEN and p53 expression in primary ovarian carcinomas: immunohistochemical study and discussion of pathogenetic mechanisms" *Int J Gynecol* Cancer 2006 Jan-Feb;16 Suppl 1:254-8.

[32] Lah TT, Duran Alonso MB, Van Noorden CJ "Antiprotease therapy in cancer: hot or not?" *Expert Opin Biol Ther* 2006 Mar;6(3):257-79.

[33] Zhang DW, Jeang KT, Lee CG "p53 negatively regulates the expression of FAT10, a gene upregulated in various cancers" *Oncogene* 2006 Feb 27 [Epub ahead of print].

[34] Li QX, Ke N, Sundarain R, Wong-Staal F ":NR4A1, 2, 3—an orphan nuclear hormone receptor family involved in cell apoptosis and carcinogenesis" *Histol Histopathol* 2006 May;21(5):533-40.

[35] Mendez O, Martin B, Sanz R, Aragues R, Moreno V, Oliva B et al. "Under expression of transcriptional regulators is common in metastatic breast cancer cells overexpressing Bcl-xL" *Carcinogenesis* 2006 Feb 20 [Epub ahead of print].

[36] Onda T, Uzawa K, Endo Y, Bickawa H, Yokoe H, Shibahara T et al. "Ubiquitous mitochondrial creatine kinase downregualted in oral squamous cell carcinoma" *Br J Cancer* 2006 Mar 13;94(5):698-709.

[37] Wang WS, Chen PM, Wang HS, Liang WY, Su Y "Matrix metalloproteinase 7 increases resistance to Fas-mediated apoptosis and is a poor prognostic factor of patients with colorectal carcinoma": *Carcinogenesis* 2006 Feb 10 [Epub ahead of print].

[38] Derka S, Vairaktaris E, Papkossa V, Vassiliou S, Acil Y, Vylliotis A et al. "Cell proliferation and apoptosis culminate in early stages of oral oncogenesis" *Oral Oncol* 2006 Feb 4 [Epub ahead of print].

[39] Parr RL, Dakubo GD, Thayer RE, McKenney K, Birch-Machin MA "Mitochondrial DNA as a potential tool for early cancer detection" *Hum Genomics* 2006;2(4):252-257.

[40] De Vincente JC, Olay S, Lequerica-Fernandez P, Sanchez-Mayoral J, Junquera LM, Fresno MF "Expression of Bcl-2 but not Bax has a prognostic significance in tongue carcinoma" *J Oral Pathol Med* 2006 Mar;35(3):140-5.

[41] Okazaki R, Moon Y, Norimura T, Eling T "Ionizing radiation enhances the expression of the nonsteroidal antiinflammatory drug-activated gene (NAG1) by increasing the expression of TP53 in human colon cancer cells" *Radiat Res* 2006 Feb;165(2):125-30.

[42] Yan B, Wang H, Peng Y, Hu Y, Wang H, Zhang X et al. "A unique roleof the DNA fragmentation factor in maintaining genomic stability" *Proc Natl Acad Sci USA*:2006 Jan31;103(5):1504-9.

[43] Sinian M, Molinolo A, Lanari C "Involvement of matrix metalloproteinase activity in hormone-induced mammary tumor regression" *Am J Pathol* 2006 Jan;168(1):270-9.

[44] Lazennec G "Estrogen receptor beta, a possible tumor suppressor involved in ovarian carcinogenesis" *Cancer Lett* 2006 Jan 18;231(2):151-7.

[45] Thomas GJ, Nystrom ML, Marshall JF "Alphavbeta6 integrin in wound healing and cancer of the oral cavity" *J Oral Pathol Med* 2006 Jan;35(1):1-10.

[46] Zhen HN, Zhang X, Hu PZ, Yang TT, Fei Z, Zhang JN et al. "Survivin expression and its relation with proliferation, apoptosis, and angiogenesis in brain gliomas" *Cancer* 2005 Dec 15;104(12):2775-83.

In: Cell Apoptosis and Cancer
Editor: Albina W. Taylor, pp. 61-74
ISBN: 978-1-60021-506-3

Chapter IV

MODULATING APOPTOSIS THRESHOLDS IN CANCER THERAPY

Christoph Herold and Matthias Ocker[1]
Department of Medicine 1, University Hospital Erlangen, Ulmenweg 18, D-91054 Erlangen, Germany

ABSTRACT

The ability to induce and execute apoptosis is essential for the maintenance and regulation of tissue and organ homeostasis. A common feature of malignant cells is their capability of evading apoptosis, which contributes to the high rate of resistance to radio- or chemotherapy of tumor cells. Although some progress has been made in inhibiting tumor growth by developing receptor-tyrosine-kinase antagonists that inhibit survival and growth promoting pathways, resistance against these and other compounds is mainly due to defects in apoptosis regulation. In contrast, low apoptotic thresholds may sensitize tumors to well tolerated and clinical established treatment regimens. Therefore, downstream targets of the apoptosis machinery, e.g. survivin, caspases or members of the bcl-2 family, seem to represent interesting new targets for cancer therapy. Modulating apoptotic thresholds can be achieved by reconstituting genetically lost or epigenetically silenced pro-apoptotic factors, e.g. by viral gene transfer or inducing re-expression with DNA methyltransferase or histone deacetylase inhibitors, or by inhibiting the activity or expression of anti-apoptotic factors. Here, RNAi-based approaches, e.g. against anti-apoptotic bcl-2, seem to hold great promise to overcome treatment resistance of a variety of malignant tumor types. Overall, restoring apoptosis sensitivity or lowering apoptotic thresholds will lead to better tolerated and more potent treatment strategies.

[1] Correspondence to: Dr. Matthias Ocker, Assistant Professor of Experimental Medicine, Department of Medicine 1, University Hospital Erlangen, Ulmenweg 18, D-91054 Erlangen, Germany, Phone: ++49-9131-8535057, Fax: ++49-9131-8535058, Matthias.Ocker@med1.imed.uni-erlangen.de

Introduction

The control of cell number, tissue integrity and homeostasis is crucial for the survival of multicellular organisms. Besides controlling proliferation of cells, the coordinated removal and killing of unwanted or unnecessary cells is a highly conserved mechanism to achieve this end. This process has been termed "apoptosis" in 1972 in analogy to the Greek for "falling off", resembling the falling of leaves from trees in autumn [1]. Apoptosis plays several critical roles in various physiologic processes ranging from embryonic development, immune maturation and response and tissue homeostasis in adult organisms. The dysregulation of apoptosis has been linked to the pathogenesis of human cancer diseases and especially to the high resistance of some cancers to conventional radio- and/or chemotherapy [2]. Recently, several genes and proteins have been identified that contribute to this dysregulation and the better understanding of how these processes themselves are regulated has lead to the development of novel therapeutic approaches, esp. applying gene technology approaches, small molecule inhibitors or epigenetic modulators.

In this chapter, we will review the basic features of apoptosis in health and cancer diseases and point out to strategies to overcome or modulate apoptotic thresholds for cancer therapy.

Basic Features of Apoptosis

Apoptosis was first recognized from a morphologic point of view by cell shrinkage, chromatin condensation and fragmentation (pyknosis and karyorrhexis), formation of apoptotic blebbing and apoptotic bodies and finally phagocytosis of apoptotic remnants by macrophages without signs of inflammation. Importantly, and in contrast to necrosis, membranes stay intact until the final stages of apoptosis execution [3]. These morphologic phenomena are caused by a series of distinct biochemical features that are executed in a coordinated fashion, e.g. the hydrolytic cleavage of cellular proteins by the caspase cascade [4], the distinct pattern of internucleosomal DNA degradation and changes in membrane architecture (i.e. expression of phosphatidylserine on the outer layers of the plasma membrane) that are necessary for the phagocytic recognition by macrophages [5, 6].

The process of programmed cell death can be divided into three phases: initiation, regulation and execution that follow each other rapidly. Initiation of apoptosis can be observed after stimulating the intrinsic or mitochondrial pathway by genotoxic damage or microtubule disruption (e.g. after radio- or chemotherapy for cancer treatment), hypoxia, hypoglycaemia, reactive oxygen species, withdrawal of growth factors or via the extrinsic pathway, i.e. stimulation of death receptors [7] (see figure 1). The phase of regulation involves the cellular decision to undergo the irreversible pathway to cell death. Here, a connection between the extrinsic and the intrinsic apoptosis induction pathways can further enhance and accelerate the final execution phase. The execution phase finally leads to the above described morphologic alterations. In the following sections, the different pathways of induction, regulation and execution of apoptosis will be highlighted in regard of their meaning for cancer development and therapy. Although the different pathways are described separately for better understanding, it is important to note that they are closely interconnected and maybe even interdependent on each other.

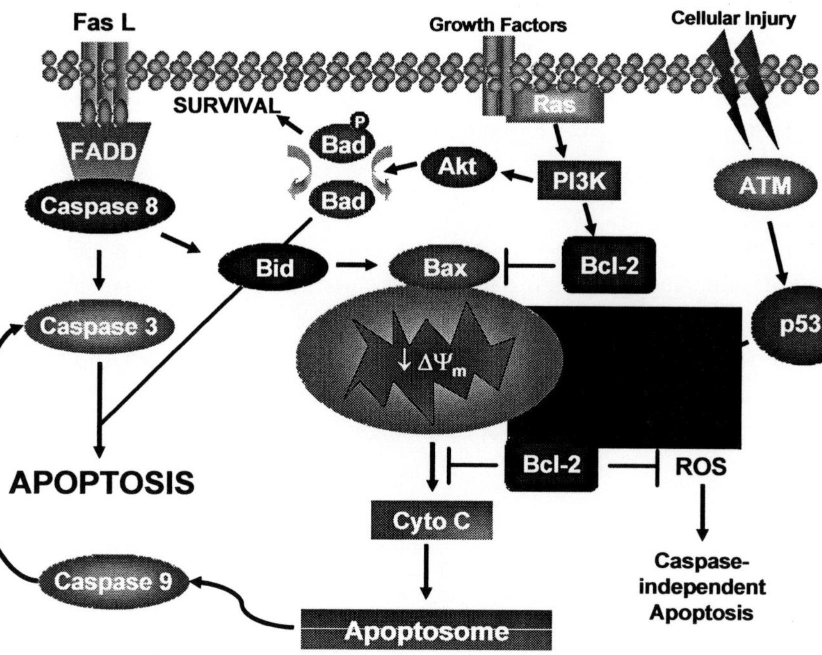

Figure 1. Schematic overview on extrinsic and intrinsic pathways of apoptosis induction. The mitochondrion represents the central decision making point for both the receptor mediated extrinsic as well as the physiologic intrinsic pathway. The release of Cytochrome *c* from the mitochondrial intermembrane space leads to the formation of the apoptosome that activates the initiator caspase 9. The receptor mediated pathway leads to the activation of intracellular death domains (FADD) that activate the initiator caspase 8. Both initiator caspases can activate the effector caspase 3 that is the final executioner of apoptosis. Activated caspase 8 can also disturb the mitochondrial integrity and lead to the loss of the mitochondrial membrane potential $\Delta\Psi_m$ with consecutive release of Cytochrome *c*. The regulation of the mitochondrial integrity is dependent on different pro- and anti-apoptotic members of the bcl-2 protein family that are differentially regulated by growth factor withdrawal or cellular stress.

Extrinsic (Receptor Mediated) Induction of Apoptosis

Death receptors represent a growing family of transmembrane receptors present on most cell types. After binding of apoptosis inducing ligands (e.g. FAS ligand, TNF-α, TNF-β, TRAIL) a cytoplasmic domain involved in protein-protein interactions called death domain (DD) is activated and recruits several adapter proteins. DD in turn binds to several inactive molecules of pro-caspase 8. Although pro-caspases are thought to represent inactive pro-forms of these proteases, they seem to posses a low intrinsic activity that allows the pro-caspase to activate other pro-caspases when brought to close proximity, e.g. after DD binding [4, 8, 9]. The thus activated initator caspase 8 then triggers a cascade of caspase activation by cleaving and activating other downstream caspases and factors leading to a destabilization of the mitochondrial membrane, e.g. cleavage of bid to t-bid [10]. The DD mediated activation of pro-caspases can be inhibited by the protein FLIP, which is commonly overexpressed in cancer cells to protect the cell from ligand induced cell death [11, 12]. Inhibition of FLIP might therefore represent a novel target for restoring apoptosis sensitivity in some tumors.

INTRINSIC (MITOCHONDRIAL) INDUCTION OF APOPTOSIS

Different cellular stress factors (e.g. irradiation, growth factor withdrawal, mitotic catastrophe etc.) lead to increased permeability of the mitochondrial membrane with subsequent release of Cytochrome *c*. In the cytosol, Cytochrome *c* then binds to the apoptosis activating factor-1 (Apaf-1) and this complex then activates the initiator caspase 9 [13]. Futhermore, antagonists of endogenous caspase inhibitors known as IAPs like Smac/Diablo are also released from mitochondria and enhance apoptotic activity [14]. The further processes resemble those described for the extrinsic pathway: activation of effector caspases and execution of cell death.

The mitochondrial pathway is strongly involved in the apoptotic response to radio- or chemotherapy. Normally, the integrity of the mitochondrial membrane is controlled by the family of bcl-2 proteins, which consists of anti-apoptotic (e.g. bcl-2, bcl-X_L, Mcl-1, Bfl-1, A1, etc.) and pro-apoptotic members (e.g. bax, bad, bak, bik, bcl-X_S, etc.) [14-17]. According to the rheostat model, the balance between pro- and anti-apoptotic factors decides about the execution of cell death [18]. Here, the ratio of bcl-2 and bax seems to be most important. Therefore, overexpression of anti-apoptotic bcl-2 or loss of pro-apoptotic bax has been shown in various human tumor diseases [19, 20].

Although the exact mechanisms of how pro- and anti-apoptotic factors decide about cell fate still remains elusive, several models have been proposed so far [21]. Pro- and anti-apoptotic bcl-2 family members share conserved stretches of amino-acid sequences that were termed bcl-2 homology (BH) domains. Among these, the BH3-containing or BH3-only proteins possess pro-apoptotic properties, e.g. bax or bak. Normally, these pro-apoptotic molecules are located in the cytoplasm in a folded configuration that hides the C-terminal mitochondrial localizing domain and renders the protein inactive. Upon a death stimulus, bax can unfold, localize to the outer mitochondrial membrane [22, 23], form multimers [24, 25], and induce mitochondrial permeabilization with subsequent loss of $\Delta\Psi_m$ and release of Cytochrome *c*. A death stimulus can e.g. be the caspase-mediated cleavage of bid to t-bid upon activation of the receptor-mediated pathway that activates the BH3-domain of t-bid [21, 26]. Furthermore, the BH3-domain can also antagonize the effect of anti-apoptotic bcl-2 family members [27, 28]. Activated bax-multimers can induce the mitochondrial permeability transition (MPT) pore, where the voltage-dependent anion channel (VDAC) seems to be an integral part [29]. According to this model, opening of VDAC then disrupts the electron gradient and leads to loss of $\Delta\Psi_m$. Another controversially debated model indicates that multimerized bax molecules can themselves form pores in the mitochondrial membrane that are large enough for releasing Cytochrome *c* [21].

Irrespective of the precise mechanism that leads to bax-mediated disruption of the mitochondrial integrity, a variety of factors are involved in upstream signaling or regulation of these molecules and several of them have been shown to be dysregulated in cancer diseases.

REGULATION OF APOPTOSIS IN CANCER

As the efficacy of cancer therapy is largely dependent on the capability to induce programmed cell death, evasion of apoptosis has been termed as one of the hallmarks of cancer [30]. Normally, proliferation and apoptosis are tightly balanced to maintain tissue homeostasis and either enhanced proliferation or impaired apoptosis lead to a net gain in cell number, i.e. tumor formation. Therefore, defects in either apoptosis induction or apoptosis execution are central for the survival of successful tumor cells (see table 1). Here, the two basic principles of apoptosis evasion (failure of apoptosis induction and failure in apoptosis execution) will be highlighted.

Table 1: Role of apoptosis deficiency in cancer [98]

Feature	Mechanism
Net increase in cell number	Cell death < cell proliferation
Longevity	Accumulation of genetic and epigenetic lesions
Proto-oncogene tolerance	Nullify pro-apoptotic effects
Immune surveillance	Resistance to immune attack
Growth factor-independence	Survival without growth factors
Angiogenesis	Resistance to hypoxia, hypoglycemia
Metastasis	Survival without attachment, resistance to anoikis
Radio-/Chemoresistance	Increased apoptotic threshold

As discussed previously, induction of apoptosis can be mediated via an extrinsic or an intrinsic pathway. Loss of expression of DD-receptors like Fas (CD95) or members of the TRAIL-family are commonly observed in human cancers and contribute to impaired overall survival [31-34]. Furthermore, inactivating point mutations in either the ligand binding domain or in the intracellular DD itself have been described [35, 36]. Another commonly observed strategy is the overexpression of so called "decoy" receptors, i.e. members of the TNF-receptor superfamily that bind pro-apoptotic ligands but do not have an intracellular DD [34, 37, 38].

Other mechanisms of apoptosis resistance in cancer focus on the intrinsic pathway and the downstream effectors of apoptosis execution. As indicated, overexpression of anti-apoptotic bcl-2 family members like bcl-2 or bcl-X_L stabilizes mitochondrial membranes or inhibits the effect of pro-apoptotic molecules like bax or t-bid [19]. Vice versa, the loss of pro-apoptotic bcl-2 family members also leads to impaired apoptosis in cancer cells. As the execution of cell death by activated caspases is normally tightly controlled, several endogenous caspase inhibitors like livin, survivin or the IAP family or mitochondrial apoptosis inhibitors like SMAC have been identified [39-44]. Another commonly denoted phenomenon in cancer is the overexpression of growth factor receptors like erb-B1 (EGFR) or erb-B2 (Her2) which can modulate the threshold of mitochondria associated apoptotic factors, e.g. by phosphorylating and thus inactivating caspase 9 by the downstream kinase AKT [45]. The phosphorylation of these proteins can also be mediated by the mitogen-activated protein kinase (MAPK) pathway.

The regulation of these pathways seems in some cancers to be dependent on the functional state of p53, which itself is commonly mutated in various cancer diseases. Therefore, mutations of p53 have been associated with impaired apoptosis due to e.g. overexpression of survivin, bcl-2 or other anti-apoptotic molecules [46-48]. Additionally, p53 has also been shown to regulate the extrinsic apoptosis cascade by influencing the expression of TRAIL- or Fas-receptors [49, 50].

STRATEGIES TO MODULATE APOPTOTIC THRESHOLDS IN CANCER THERAPY

The increasing understanding of regulatory processes involved in apoptosis induction and execution has offered the prospect of novel therapeutic approaches. In fact, the selective induction of apoptosis via the extrinsic pathway has been pursued in different cancers using recombinant antibodies [51, 52]. Yet, the application of Fas- or TRAIL-receptor specific antibodies or other activators like recombinant human tumor necrosis factor (rhTNF) or soluble Fas ligands (FasL/CD95L) showed significant levels of toxicity in various organs up to sepsis like syndromes [53, 54]. Gene therapy approaches using adenoviral vectors for re-expressing Fas or FasL have not shown convincing results overall [51].

As indicated, the activity of apoptotic pathways is usually under control of pro-survival (e.g. growth factor signaling) pathways. Here, the development of novel small molecule inhibitors of different kinases involved in this signaling has lead to better treatment responses due to shifts in apoptotic thresholds, too. Yet, due to the novelty of these treatments, e.g. targeting EGFR, MAPK, AKT or PI3K pathways, no final conclusion on the specifity and efficacy can be drawn.

Different strategies have been followed to modulate the bax/bcl-2 ratio in tumor cells. Antisense oligonucleotides against bcl-X_L or bcl-2 have been successfully used in various cancer cell models *in vitro*. However, preclinical and clinical studies did not demonstrate convincing anti-cancer efficacies due to unspecific effects (e.g. complement activation, thrombocytopenia, cytokine release, hepatotoxicity) and the need to introduce chemical modifications (e.g. locked nucleic acids) to stabilize these molecules [55-64]. Recently, a novel and more specific approach to suppress the function of oncogenes has been described in *C. elegans* and mammals: RNA interference (RNAi) [65-67]. Briefly, short double-stranded RNA molecules (siRNA) with homology to the mRNA sequence of a target gene are processed enzymatically and lead to the cleavage and suppression of the target mRNA with high specificity [68, 69]. We applied this technology to silence the expression of anti-apoptotic bcl-2 in a pancreatic cancer model, which is otherwise highly resistant to chemotherapy. In this setting, the knockdown of bcl-2 by siRNA lead to the induction of apoptosis and reduced cell proliferation *in vitro* and in a xenograft model *in vivo* [70]. This effect was mediated by a loss of bcl-2 and an enhanced destabilization of the mitochondrial transmembrane potential with subsequent loss of $\Delta\Psi_m$ (Figure 2). We next applied the cytidine analogue Gemcitabine, which has become the gold standard in the treatment of advanced pancreatic cancer, after silencing bcl-2. In cells with silenced bcl-2, Gemcitabine was able to induce cell death even at concentrations 10-times below the effective threshold of Gemcitabine monotherapy (Figure 3). This approach seems to be highly specific, as control experiments did not show any efficacy of this low dose of Gemcitabine. Although the broad application of RNAi based therapies is still limited, these experiments have shown that modulating the apoptotic bax/bcl-2 threshold restores sensitivity to chemotherapy even at very low doses. This indicates that with this adjunctive treatment either an enhanced efficacy or less undesired side effects could be achieved in patients who may otherwise be untreatable at all. This approach has also been successfully applied in other model system [71-79].

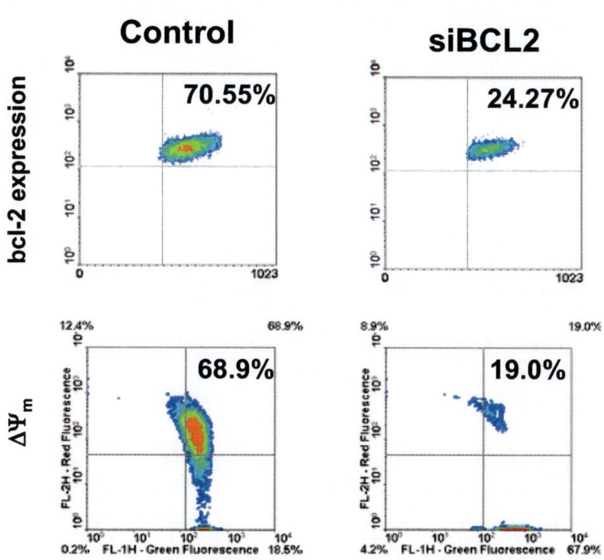

Figure 2. siRNA-mediated knockdown of anti-apoptotic bcl-2 [70]. Pancreatic cancer cells were transfected with siRNA against bcl-2 (siBCL2) and the expression of bcl-2 was determined by flow cytometry using a FITC-labeled anti-bcl-2 antibody. In untreated controls, 70.55% of cells expressed bcl-2, while in treated cells, only 24.27% showed a positive staining. This correlated with the loss of $\Delta\Psi_m$ as evidenced after JC-1 staining in flow cytometry. Here, 68.9% of untreated controls exhibited intact mitochondria, while transfection with siBCL2 reduced this level to 19%.

Recently, so called epigenetic events have attracted attention as novel cancer therapeutics [80, 81]. Among these, small molecule inhibitiors of histone deacetylases (HDAC), are currently undergoing clinical trials for different tumor diseases [82-84]. These compounds have been shown to modulate the bax/bcl-2 ratio of cancer cells as well as the expression of TRAIL- and other death receptors [85-90]. We and others have shown that HDAC inhibitors can restore sensitivity to radio- or chemotherapy in otherwise resistant tumor entities at low molecular concentrations [91-95]. These effects are explained by either a transcriptional regulation of apoptosis relevant target genes like bcl-2, a suppression of multidrug resistance proteins or a halt in cell cycle progression, e.g. by upregulation of $p21^{cip1/waf1}$, which allows conventional cytostatic or cytotoxic drugs to exert their growth-inhibitory effect for a longer time period [91].

Another principle of epigenetic regulation is the hypermethylation of DNA elements [81]. Here, novel compounds like Zebularine also show promising results in preclinical settings of human cancer diseases and the combination of classical cytotoxic agents, targeted therapies and epigenetic modulators can provide potent, specific and well tolerable treatment options [81, 96].

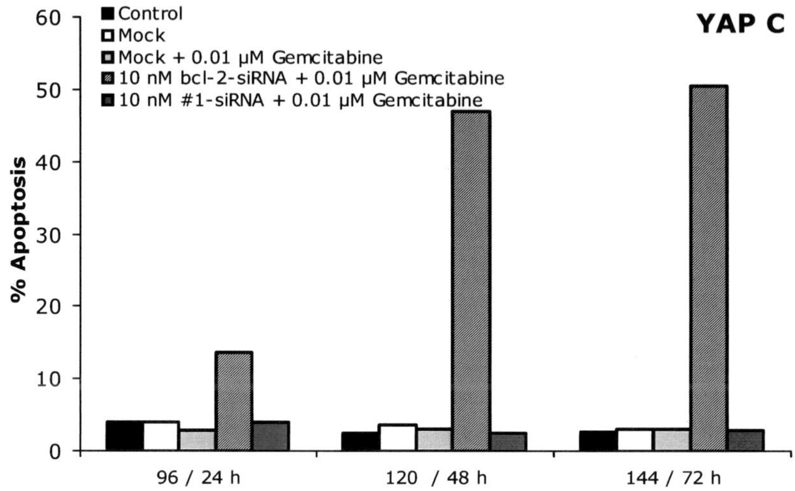

Figure 3. Loss of bcl-2 sensitizes human pancreatic cancer cells for Gemcitabine treatment. Cells were pre-transfected with siBCL2 and then incubated with Gemcitabine at an ineffective concentration. Apoptosis induction was determined by flow cytometry after propidium ioded staining. Only in cells with silenced bcl-2, significant levels of apoptosis were reached, while in controls (mock transfected or control siRNA #1), no changes compared to control were seen.

CONCLUSION

The increasing understanding of apoptosis pathways, its regulation and dysregulation in cancer leads to the generation of a variety of novel treatment approaches. Especially epigenetic events enter the focus of current research, as the discussion of the role of so-called cancer stem cells seems to lead to a paradigm shift in the understanding of cancer biology and treatment [97]. Targeted approaches, using RNAi based strategies or specific tyrosine kinase inhibitors, as well as the development of HDAC inhibitors will soon enter clinical oncology and an increased overall survival and clinical benefit is expected for otherwise untreatable patients.

REFERENCES

[1] Kerr, J.F., A.H. Wyllie, and A.R. Currie, Apoptosis: a basic biological phenomenon with wide-ranging implications in tissue kinetics. *Br J Cancer,* 1972. 26(4): p. 239-57.

[2] Reed, J.C., Dysregulation of apoptosis in cancer. *J Clin Oncol,* 1999. 17(9): p. 2941-53.

[3] Kumar, V., A.K. Abbas, and N. Fausto, *Cellular Adaptions, Cell Injury, and Cell Death, in Pathologic Basis of Disease,* V. Kumar, A.K. Abbas, and N. Fausto, Editors. 2005, Elsevier Saunders: Philadelphia. p. 3-46.

[4] Shi, Y., Caspase activation: revisiting the induced proximity model. *Cell,* 2004. 117(7): p. 855-8.

[5] Schultz, D.R. and W.J. Harrington, Jr., Apoptosis: programmed cell death at a molecular level. *Semin Arthritis Rheum,* 2003. 32(6): p. 345-69.

[6] Kasof, G., et al., Overview: A Matter of Life and Death, in *Signalling Pathways in Apoptosis,* D. Watters and M. Lavin, Editors. 1999, Harwood Academic Publishers: ,Amsterdam. p. 1-28.

[7] Fennell, D.A., Apoptosis: molecular physiology and significance for cancer therapeutics, in *Introduction to the Cellular and Molecular Biology of Cancer,* M. Knowles and P. Selby, Editors. 2005, Oxford University Press: Oxford. p. 210-28.

[8] Salvesen, G.S. and V.M. Dixit, Caspase activation: the induced-proximity model. *Proc Natl Acad Sci U S A,* 1999. 96(20): p. 10964-7.

[9] Srinivasula, S.M., et al., Autoactivation of procaspase-9 by Apaf-1-mediated oligomerization. *Mol Cell,* 1998. 1(7): p. 949-57.

[10] Esposti, M.D., The roles of Bid. *Apoptosis,* 2002. 7(5): p. 433-40.

[11] Ryu, B.K., et al., Increased expression of cFLIP(L) in colonic adenocarcinoma. *J Pathol,* 2001. 194(1): p. 15-9.

[12] Zhou, X.D., et al., Overexpression of cellular FLICE-inhibitory protein (FLIP) in gastric adenocarcinoma. *Clin Sci (Lond),* 2004. 106(4): p. 397-405.

[13] Schafer, Z.T. and S. Kornbluth, The apoptosome: physiological, developmental, and pathological modes of regulation. *Dev Cell,* 2006. 10(5): p. 549-61.

[14] Brenner, C. and G. Kroemer, The Mitochondrion: Decisive for Cell Death Control?, in *Signalling Pathways in Apoptosis,* D. Watters and M. Lavin, Editors. 1999, Harwood Academic Publishers: Amsterdam. p. 207-26.

[15] Thomadaki, H. and A. Scorilas, BCL2 family of apoptosis-related genes: functions and clinical implications in cancer. *Crit Rev Clin Lab Sci,* 2006. 43(1): p. 1-67.

[16] Armstrong, J.S., Mitochondrial membrane permeabilization: the sine qua non for cell death. *Bioessays,* 2006. 28(3): p. 253-60.

[17] Mohamad, N., et al., Mitochondrial apoptotic pathways. *Biocell,* 2005. 29(2): p. 149-61.

[18] Korsmeyer, S.J., et al., Bcl-2/Bax: a rheostat that regulates an anti-oxidant pathway and cell death. *Semin Cancer Biol,* 1993. 4(6): p. 327-32.

[19] Reed, J.C., Bcl-2 family proteins: regulators of apoptosis and chemoresistance in hematologic malignancies. *Semin Hematol,* 1997. 34(4 Suppl 5): p. 9-19.

[20] Reed, J.C., et al., BCL-2 family proteins: regulators of cell death involved in the pathogenesis of cancer and resistance to therapy. *J Cell Biochem,* 1996. 60(1): p. 23-32.

[21] Sharpe, J.C., D. Arnoult, and R.J. Youle, Control of mitochondrial permeability by Bcl-2 family members. *Biochim Biophys Acta,* 2004. 1644(2-3): p. 107-13.

[22] Wolter, K.G., et al., Movement of Bax from the cytosol to mitochondria during apoptosis. *J Cell Biol,* 1997. 139(5): p. 1281-92.

[23] Hsu, Y.T., K.G. Wolter, and R.J. Youle, Cytosol-to-membrane redistribution of Bax and Bcl-X(L) during apoptosis. *Proc Natl Acad Sci U S A,* 1997. 94(8): p. 3668-72.

[24] Tan, Y.J., W. Beerheide, and A.E. Ting, Biophysical characterization of the oligomeric state of Bax and its complex formation with Bcl-XL. *Biochem Biophys Res Commun,* 1999. 255(2): p. 334-9.

[25] Antonsson, B., et al., Bax is present as a high molecular weight oligomer/complex in the mitochondrial membrane of apoptotic cells. *J Biol Chem,* 2001. 276(15): p. 11615-23.

[26] Yin, X.M., Bid, a BH3-only multi-functional molecule, is at the cross road of life and death. *Gene*, 2006. 369: p. 7-19.

[27] Kelekar, A. and C.B. Thompson, Bcl-2-family proteins: the role of the BH3 domain in apoptosis. *Trends Cell Biol*, 1998. 8(8): p. 324-30.

[28] Huang, D.C. and A. Strasser, BH3-Only proteins-essential initiators of apoptotic cell death. *Cell,* 2000. 103(6): p. 839-42.

[29] Jacotot, E., et al., Mitochondrial membrane permeabilization during the apoptotic process. *Ann N Y Acad Sci,* 1999. 887: p. 18-30.

[30] Hanahan, D. and R.A. Weinberg, The hallmarks of cancer. *Cell,* 2000. 100(1): p. 57-70.

[31] Wajant, H., K. Pfizenmaier, and P. Scheurich, TNF-related apoptosis inducing ligand (TRAIL) and its receptors in tumor surveillance and cancer therapy. *Apoptosis*, 2002. 7(5): p. 449-59.

[32] Zhang, L. and B. Fang, Mechanisms of resistance to TRAIL-induced apoptosis in cancer. *Cancer Gene Ther,* 2005. 12(3): p. 228-37.

[33] Wajant, H., CD95L/FasL and TRAIL in tumour surveillance and cancer therapy. *Cancer Treat Res*, 2006. 130: p. 141-65.

[34] Bouralexis, S., D.M. Findlay, and A. Evdokiou, Death to the bad guys: targeting cancer via Apo2L/TRAIL. *Apoptosis*, 2005. 10(1): p. 35-51.

[35] Takahashi, H., et al., FAS death domain deletions and cellular FADD-like interleukin 1beta converting enzyme inhibitory protein (long) overexpression: alternative mechanisms for deregulating the extrinsic apoptotic pathway in diffuse large B-cell lymphoma subtypes. Clin Cancer Res, 2006. 12(11 Pt 1): p. 3265-71.

[36] Shen, L., et al., Frequent deletion of Fas gene sequences encoding death and transmembrane domains in nasal natural killer/T-cell lymphoma. *Am J Pathol*, 2002. 161(6): p. 2123-31.

[37] LeBlanc, H.N. and A. Ashkenazi, Apo2L/TRAIL and its death and decoy receptors. *Cell Death Differ*, 2003. 10(1): p. 66-75.

[38] Ozoren, N. and W.S. El-Deiry, Cell surface Death Receptor signaling in normal and cancer cells. *Semin Cancer Biol*, 2003. 13(2): p. 135-47.

[39] Deveraux, Q.L., et al., Endogenous inhibitors of caspases. *J Clin Immunol,* 1999. 19(6): p. 388-98.

[40] Deveraux, Q.L. and J.C. Reed, IAP family proteins--suppressors of apoptosis. *Genes Dev*, 1999. 13(3): p. 239-52.

[41] Roy, N., et al., The c-IAP-1 and c-IAP-2 proteins are direct inhibitors of specific caspases. *Embo J,* 1997. 16(23): p. 6914-25.

[42] Deveraux, Q.L., et al., X-linked IAP is a direct inhibitor of cell-death proteases. *Nature,* 1997. 388(6639): p. 300-4.

[43] LaCasse, E.C., et al., The inhibitors of apoptosis (IAPs) and their emerging role in cancer. *Oncogene*, 1998. 17(25): p. 3247-59.

[44] Ambrosini, G., C. Adida, and D.C. Altieri, A novel anti-apoptosis gene, survivin, expressed in cancer and lymphoma. *Nat Med*, 1997. 3(8): p. 917-21.

[45] Cardone, M.H., et al., Regulation of cell death protease caspase-9 by phosphorylation. *Science*, 1998. 282(5392): p. 1318-21.

[46] Vegran, F., et al., Association of p53 gene alterations with the expression of antiapoptotic survivin splice variants in breast cancer. *Oncogene,* 2006.

[47] Tsutsui, S., et al., Bcl-2 protein expression is associated with p27 and p53 protein expressions and MIB-1 counts in breast cancer. *BMC Cancer*, 2006. 6(1): p. 187.

[48] Yu, J. and L. Zhang, The transcriptional targets of p53 in apoptosis control. *Biochem Biophys Res Commun*, 2005. 331(3): p. 851-8.

[49] Sheikh, M.S., et al., p53-dependent and -independent regulation of the death receptor KILLER/DR5 gene expression in response to genotoxic stress and tumor necrosis factor alpha. *Cancer Res,* 1998. 58(8): p. 1593-8.

[50] Liu, X., et al., p53 upregulates death receptor 4 expression through an intronic p53 binding site. *Cancer Res,* 2004. 64(15): p. 5078-83.

[51] Timmer, T., E.G. de Vries, and S. de Jong, Fas receptor-mediated apoptosis: a clinical application? *J Pathol,* 2002. 196(2): p. 125-34.

[52] de Jong, S., et al., Death receptor ligands, in particular TRAIL, to overcome drug resistance. *Cancer Metastasis Rev*, 2001. 20(1-2): p. 51-6.

[53] Kakinuma, C., et al., Acute toxicity of an anti-Fas antibody in mice. *Toxicol Pathol,* 1999. 27(4): p. 412-20.

[54] Abe, R., et al., Toxic epidermal necrolysis and Stevens-Johnson syndrome are induced by soluble Fas ligand. *Am J Pathol,* 2003. 162(5): p. 1515-20.

[55] Waters, J.S., et al., Phase I clinical and pharmacokinetic study of bcl-2 antisense oligonucleotide therapy in patients with non-Hodgkin's lymphoma. *J Clin Oncol*, 2000. 18(9): p. 1812-23.

[56] Cotter, F.E., J. Waters, and D. Cunningham, Human Bcl-2 antisense therapy for lymphomas. *Biochim Biophys Acta,* 1999. 1489(1): p. 97-106.

[57] Pepper, C., et al., Antisense-mediated suppression of Bcl-2 highlights its pivotal role in failed apoptosis in B-cell chronic lymphocytic leukaemia. *Br J Haematol*, 1999. 107(3): p. 611-5.

[58] Chi, K.C., et al., Effects of Bcl-2 modulation with G3139 antisense oligonucleotide on human breast cancer cells are independent of inherent Bcl-2 protein expression. *Breast Cancer Res Treat*, 2000. 63(3): p. 199-212.

[59] Wacheck, V., et al., Bcl-2 antisense oligonucleotides chemosensitize human gastric cancer in a SCID mouse xenotransplantation model. *J Mol Med,* 2001. 79(10): p. 587-93.

[60] Jansen, B., et al., Chemosensitisation of malignant melanoma by BCL2 antisense therapy. *Lancet*, 2000. 356(9243): p. 1728-33.

[61] Lebedeva, I. and C.A. Stein, Antisense oligonucleotides: promise and reality. *Annu Rev Pharmacol Toxicol,* 2001. 41: p. 403-19.

[62] Gautschi, O., et al., Activity of a novel bcl-2/bcl-xL-bispecific antisense oligonucleotide against tumors of diverse histologic origins. *J Natl Cancer Inst,* 2001. 93(6): p. 463-71.

[63] Hopkins-Donaldson, S., et al., Induction of apoptosis and chemosensitization of mesothelioma cells by Bcl-2 and Bcl-xL antisense treatment. *Int J Cancer,* 2003. 106(2): p. 160-6.

[64] Stahel, R.A. and U. Zangemeister-Wittke, Antisense oligonucleotides for cancer therapy-an overview. *Lung Cancer,* 2003. 41 Suppl 1: p. S81-8.

[65] Fire, A., et al., Potent and specific genetic interference by double-stranded RNA in Caenorhabditis elegans. *Nature,* 1998. 391(6669): p. 806-11.

[66] Elbashir, S.M., et al., Duplexes of 21-nucleotide RNAs mediate RNA interference in cultured mammalian cells. Nature, 2001. 411(6836): p. 494-8.

[67] Elbashir, S.M., W. Lendeckel, and T. Tuschl, RNA interference is mediated by 21- and 22-nucleotide RNAs. *Genes Dev,* 2001. 15(2): p. 188-200.

[68] Meister, G. and T. Tuschl, Mechanisms of gene silencing by double-stranded RNA. *Nature,* 2004. 431(7006): p. 343-9.

[69] Tuschl, T., RNA interference and small interfering RNAs. *Chembiochem*, 2001. 2(4): p. 239-45.

[70] Ocker, M., et al., Variants of bcl-2 specific siRNA for silencing antiapoptotic bcl-2 in pancreatic cancer. *Gut*, 2005. 54(9): p. 1298-308.

[71] Rödel, F., et al., Survivin as a radioresistance factor, and prognostic and therapeutic target for radiotherapy in rectal cancer. *Cancer Res,* 2005. 65(11): p. 4881-7.

[72] Ling, X. and F. Li, Silencing of antiapoptotic survivin gene by multiple approaches of RNA interference technology. *Biotechniques*, 2004. 36(3): p. 450-4, 456-60.

[73] Chawla-Sarkar, M., et al., Downregulation of Bcl-2, FLIP or IAPs (XIAP and survivin) by siRNAs sensitizes resistant melanoma cells to Apo2L/TRAIL-induced apoptosis. *Cell Death Differ*, 2004. 11(8): p. 915-23.

[74] Collis, S.J., et al., Enhanced radiation and chemotherapy-mediated cell killing of human cancer cells by small inhibitory RNA silencing of DNA repair factors. *Cancer Res*, 2003. 63(7): p. 1550-4.

[75] Kosciolek, B.A., et al., Inhibition of telomerase activity in human cancer cells by RNA interference. *Mol Cancer Ther*, 2003. 2(3): p. 209-16.

[76] Salvi, A., et al., Small interfering RNA urokinase silencing inhibits invasion and migration of human hepatocellular carcinoma cells. *Mol Cancer Ther,* 2004. 3(6): p. 671-8.

[77] Schmitz, J.C., T.M. Chen, and E. Chu, Small interfering double-stranded RNAs as therapeutic molecules to restore chemosensitivity to thymidylate synthase inhibitor compounds. *Cancer Res,* 2004. 64(4): p. 1431-5.

[78] Takei, Y., et al., A small interfering RNA targeting vascular endothelial growth factor as cancer therapeutics. *Cancer Res,* 2004. 64(10): p. 3365-70.

[79] Yano, J., et al., Antitumor Activity of Small Interfering RNA/Cationic Liposome Complex in Mouse Models of Cancer. *Clin Cancer Res,* 2004. 10(22): p. 7721-6.

[80] Liu, T., et al., Histone deacetylase inhibitors: multifunctional anticancer agents. *Cancer Treat Rev,* 2006. 32(3): p. 157-65.

[81] Esteller, M., CpG island methylation and histone modifications: biology and clinical significance. Ernst Schering *Res Found Workshop,* 2006(57): p. 115-26.

[82] Kuendgen, A., et al., The histone deacetylase (HDAC) inhibitor valproic acid as monotherapy or in combination with all-trans retinoic acid in patients with acute myeloid leukemia. *Cancer,* 2006. 106(1): p. 112-9.

[83] Kelly, W.K., et al., Phase I study of an oral histone deacetylase inhibitor, suberoylanilide hydroxamic acid, in patients with advanced cancer. *J Clin Oncol,* 2005. 23(17): p. 3923-31.

[84] Ryan, Q.C., et al., Phase I and pharmacokinetic study of MS-275, a histone deacetylase inhibitor, in patients with advanced and refractory solid tumors or lymphoma. *J Clin Oncol,* 2005. 23(17): p. 3912-22.

[85] Wang, S., et al., Activation of Mitochondrial Pathway is Crucial for Tumor Selective Induction of Apoptosis by LAQ824. *Cell Cycle,* 2006. 5(15).

[86] Ziauddin, M.F., et al., Valproic Acid, an Antiepileptic Drug with Histone Deacetylase Inhibitory Activity, Potentiates the Cytotoxic Effect of Apo2L/TRAIL on Cultured Thoracic Cancer Cells through Mitochondria-Dependent Caspase Activation. *Neoplasia,* 2006. 8(6): p. 446-57.

[87] Pathil, A., et al., HDAC inhibitor treatment of hepatoma cells induces both TRAIL-independent apoptosis and restoration of sensitivity to TRAIL. *Hepatology,* 2006. 43(3): p. 425-34.

[88] Schuchmann, M., et al., Histone deacetylase inhibition by valproic acid down-regulates c-FLIP/CASH and sensitizes hepatoma cells towards CD95- and TRAIL receptor-mediated apoptosis and chemotherapy. *Oncol Rep,* 2006. 15(1): p. 227-30.

[89] Ganslmayer, M., et al., A quadruple therapy synergistically blocks proliferation and promotes apoptosis of hepatoma cells. *Oncol Rep,* 2004. 11(5): p. 943-50.

[90] Herold, C., et al., The histone-deacetylase inhibitor Trichostatin A blocks proliferation and triggers apoptotic programs in hepatoma cells. *J Hepatol,* 2002. 36(2): p. 233-40.

[91] Ocker, M., et al., The histone-deacetylase inhibitor SAHA potentiates proapoptotic effects of 5-fluorouracil and irinotecan in hepatoma cells. *J Cancer Res Clin Oncol,* 2005. 131(6): p. 385-94.

[92] Chinnaiyan, P., et al., Modulation of radiation response by histone deacetylase inhibition. *Int J Radiat Oncol Biol Phys,* 2005. 62(1): p. 223-9.

[93] Zhang, Y., et al., Enhancement of radiation sensitivity of human squamous carcinoma cells by histone deacetylase inhibitors. *Radiat Res,* 2004. 161(6): p. 667-74.

[94] Maiso, P., et al., The histone deacetylase inhibitor LBH589 is a potent antimyeloma agent that overcomes drug resistance. *Cancer Res,* 2006. 66(11): p. 5781-9.

[95] Piacentini, P., et al., Trichostatin A enhances the response of chemotherapeutic agents in inhibiting pancreatic cancer cell proliferation. *Virchows Arch,* 2006. 448(6): p. 797-804.

[96] Neureiter, D., et al., Apoptosis, proliferation and differentiation patterns are influenced by Zebularine and SAHA in pancreatic cancer models. *Scand J Gastroenterol,* 2006. in press.

[97] Neureiter, D., C. Herold, and M. Ocker, Gastrointestinal cancer - only a deregulation of stem cell differentiation? *Int J Mol Med,* 2006. 17(3): p. 483-9.

[98] Reed, J.C., Apoptosis and Cancer, in *Cancer Medicine 7*, D.W. Kufe, et al., Editors. 2006, BC Decker Inc.: Hamilton. p. 41-52.

In: Cell Apoptosis and Cancer
Editor: Albina W. Taylor, pp. 75-85

ISBN: 978-1-60021-506-3

Chapter V

APOPTOSIS IN THE BATTLE FIELD OF THE MELANOMA TUMOR MICROENVIRONMENT

Yuru Meng and Thomas F. Gajewski*

Departments of Pathology and Medicine, Committees on Immunology and Cancer Biology, University of Chicago, USA.

ABSTRACT

Tumors frequently contain activated $CD8^+$ T cells that have the potential to be cytolytic. However, spontaneous immune-mediated rejection of cancers occurs infrequently. Although there are several potential explanations for this observation, one leading mechanism may be the ultimate selection of tumor cells intrinsically resistant to immune-mediated apoptosis. Collectively, the integrated contribution of pro- and anti-apoptotic signaling factors in each individual tumor cell may decide its fate and ability to survive the attempts at killing by $CD8^+$ T cells. In this article, we will review how melanoma cells may be resistant to triggers of apoptosis. Overexpression of Bcl-2 family members, IAP family members, and survivin is commonly seen in melanoma. Resistance to cytotoxic granule-mediated lysis may be mediated through expression of Serpins, and escape from FasL-induced apoptosis may occur through downregulation of Fas and alterations in Fas signaling components. Finally, recent observations regarding the oncogenic events in melanomagenesis, such as activation of Notch signaling, have provided new insights into the potential biochemical regulation of an anti-apoptotic program.

* Correspondence concerning this article should be addressed to Yuru Meng / Thomas F. Gajewski, University of Chicago, 5841 S. Maryland Ave., MC2115, Chicago, IL 60637. Phone: 773-702-4601; Fax: 773-702-3163; email: ruthmwk@yahoo.com or tgajewsk@medicine.bsd.uchicago.edu.

INTRODUCTION

Cellular immunity has tremendous potential as a cancer therapeutic [1,2]. Cancer patients can generate tumor antigen-specific immune responses, either spontaneously or following vaccination or adoptive transfer, and effector T lymphocytes thus generated can in many cases migrate into the tumor microenvironment, the battle field [3]. Theoretically, these effector tumor infiltrating lymphocytes (TILs) have the potential to produce IFN-γ, TNF-α, and other inflammatory cytokines. They also can contain cytotoxic granules and inducibly express membrane-bound Fas ligand (FasL) (**Figure 1a**), through which they can kill tumor cells via apoptosis. However, despite the potential of the immune system to destroy tumor cells, spontaneous immune-mediated rejection of cancers is an uncommon event. Several potential explanations for this paradox in melanoma have been described. Immunosuppressive events within the tumor microenvironment have recently received much attention. T cell anergy due to insufficient costimulation [4,5], extrinsic suppression by $CD4^+CD25^+FoxP3^+$ regulatory T cells [6-8], ligation of inhibitory receptors such as CTLA-4 and PD-1 on the surface of tumor-specific T cells [9-12], and metabolic regulation by indoleamine-2,3-dioxygenase or arginase [3,13] are all being investigated as mechanisms of immune escape, a topic which has recently been reviewed [3]. However, even if each of these dominant immune escape mechanisms is successfully countered, the ultimate killing event by $CD8^+$ effector T cells relies on tumor cell susceptibility to apoptosis. Ultimately, selection of tumor cells intrinsically resistant to immune-mediated apoptosis will likely occur. Understanding the details of tumor cell resistance to apoptosis will likely represent an important final piece of the immunotherapy puzzle. The focus of this review is on how melanoma cells may resist CTL-mediated apoptosis during their close interaction in the tumor microenvironment (**Figure 1b**).

Resistance to Cytotoxic Granules

Cytotoxic granules play a major role in immune effector function [14,15]. The intracellular delivery of serine proteases, such as granzyme B (GrB), has been reported to cleave the inhibitor of caspase-activated DNase (ICAD) and numerous caspases then induce target cell apoptosis [16-18]. It used to be widely accepted that GrB entry into target cells relied on pore-formation by perforin molecules during the receptor-triggered secretory process. Recent work has suggested that GrB may enter into the target cells by endocytosis [17,19,20], or binding a proposed GrB receptor encoded by a gene located at human chromosome 6q which associated with mutations in different tumors including melanoma [21,22]. This genetic link may provide melanoma cells an inherent resistance to GrB. However, Serpins, a family of serine protease inhibitors, can efficiently and irreversibly inactivate GrB and block GrB/perforin-induced apoptosis [23,24]. The serpin PI-9 has been found to be constitutively expressed in human melanoma cells which is associated with resistance to CTL-mediated apoptosis [25]. Manipulation of the murine equivalent Spi6 can determine whether melanoma target cells are killed or not by specific $CD8^+$ CTL in vitro

[26,27]. Thus, countering the expression or function of serpins in tumor cells may be a critical step in rendering tumors more susceptible to immune-mediated destruction.

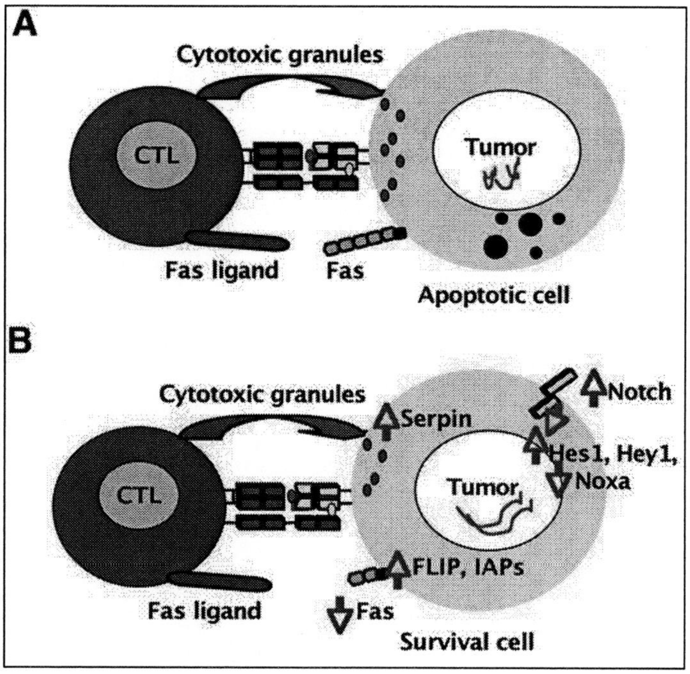

Figure 1. Interactions between tumor cells and anti-tumor CTL. A. activated anti-tumor CTLs recognize tumor cells and release cytotoxic granules which enter the target cell. Activated anti-tumor CTLs may also upregulate FasL which kill both cognate and non-cognate tumor cell. B. Serpins expressed by tumor cells may attenuate the pro-apoptotic effect of GrB-containing granules released by CTLs. Tumor cells may also down-regulate Fas expression on the surface, and increase expression of FLIP and IAPs to escape FasL-mediated apoptosis. Notch upregulation and constitutively activation in tumor cells may lead to decreased expression of NOXA, which diminishes apoptotic potential.

Resistance to Ligation of Fas and TRAIL

Ligation of Fas on tumor cells is a second major mechanism of immune-mediated killing. Genetic variants in the Fas gene have been shown not only contribute to the etiology of malignant melanoma [28], but elevated levels of soluble Fas (sFas) in melanoma explants and in the serum of melanoma patients is also associated with poor prognosis [29-31]. sFas can act as a decoy for FasL to competitively inhibit Fas-mediated apoptosis [32]. Furthermore, many cell lines derived from melanoma metastases have lower levels or no Fas expression [33], and downregulation Fas receptors on the surface of cancer cells has been suggested to protect some tumor cells from Fas-mediated death [34]. In the tumor microenvironment, cancer cells can rapid translocate Fas receptors to the cytosol from the membrane anchored to the actin cytoskeleton, thus avoiding ligation. Reduction of Fas receptor expression on the cell surface can efficiently inhibit its clustering and internalization during early apoptosis induced by anti-Fas antibody [34,35].

Even in tumor cells that retain Fas expression resistance to CD95-mediated apoptosis has been described. FasL ligation on these tumor cells can lead to activation of NF-κB and Erk1/2 signaling pathways, resulting in increased motility and invasiveness rather than death [36]. Expression of inhibitors of Fas-mediated apoptosis by tumor cells, such as FLIP and IAPs, also contributes to resistance [37-39]. FLIP (FLICE inhibitory protein) has been reported to be elevated in melanoma cell lines and malignant melanoma tumors. Interestingly, phosphorylation of FLIP occurs following exposure to Cisplatin which correlates with apoptosis in melanoma cells [40,41].

Like FasL, TRAIL is another apoptosis-inducing ligand that has potential for killing melanoma tumor cells. TRAIL binds the death-inducing receptors, TRAIL-R1 (DR4) and TRAIL-R2 (DR5), resulting in receptor cross-linking, caspase activation, and apoptosis [42-44]. However, studies have shown that resistance in the intrinsic apoptotic pathway also plays a critical role in regulating sensitivity of melanoma cells to TRAIL-induced death [45,46]. The susceptibility of melanoma cells to FasL- and/or TRAIL-induced apoptosis heavily depends on the signaling strength of pro- and anti-apoptotic signaling factors [47].

General Anti-Apoptotic Mechanisms

The apoptotic process is tightly controlled by several well-defined factors. Dysregulated apoptosis is one contributing factor to cancer development [48]. In particular, the delicate balance of pro-apoptotic Bcl-2 family proteins (e.g. Bax, Bad, Bid) and anti-apoptotic Bcl-2 family members (e.g. Bcl-2 and Bcl-X_L) regulates the intrinsic apoptotic pathway through transcription, proteolytic cleavage, or phosphorylation. Bcl-2 has been shown to lead to prolonged survival of cells by preventing apoptosis in a variety of physiological, pathological and pharmacological conditions [49]. Upregulated Bcl-2 expression in melanoma may confer tumor resistance to chemotherapy and irradiation therapy. High levels of Bcl-2 expression is associated with progression of malignant melanoma and poor response to chemotherapy [50-52]. However, there are some inconsistencies in the literature as to whether Bcl-2 represents the major resistance mechanism [53]. Other Bcl-2 family members are also differentially expressed in human melanocytes, melanoma metastases and melanoma cell lines [54,55]. Bcl-X_L expression also has been shown to be increased in the progression of primary melanoma to melanoma metastases [55]. Based on these observations, bispecific antisense oligonucleotides toward both Bcl-2 and Bcl- X_L have been investigated and have been shown to induce apoptosis of melanoma cells in vitro [56].

IAPs (Inhibitors of apoptosis proteins) prevent activation of caspases and also may render melanoma cells resistant to apoptotic inducers [57,58]. One of the IAP family members, survivin, appears to regulate mitochondrial apoptosis upstream of caspase activation and may also protect cells against caspase-independent apoptosis [59-61]. As a multifunctional protein regulating cell viability at the level of mitotic and stress response checkpoints, survivin is highly expressed in various cancers [62-66]. Its expression has been correlated with tumor aggression and poor prognosis for many cancers including melanoma [67]. Reducing survivin expression or phosphorylation by siRNAs or point mutation can sensitize melanoma cells to apoptosis [65,68-72].

Activation of Notch Signaling

Recent investigation has turned attention to oncogenic pathways in melanoma that may favor resistance to apoptosis. The Notch signaling pathway was discovered and named through genetic studies in a fruit fly with the phenotype of notched wings. The Notch receptor is a single transmembrane protein containing an extracellular ligand-binding domain and nuclear CSL binding domain [71,72]. In the early 1990s, four mammalian Notch receptor homologs (Notch 1–4) were identified. Since the Notch genes are located in regions of chromosomal translocations in cancer [73], associations between altered Notch signaling and malignant transformation have emerged [74,75]. Perturbations in Notch signaling have been linked with numerous tumor types including melanoma, providing a rationale to explore Notch signaling as a therapeutic target in cancers [71,72,75,76]. Our own unpublished data have confirmed the expression of active Notch family members in the majority of human melanoma cell lines tested (Meng and Gajewski, unpublished observations). Recent work from Nickoloff and colleagues has suggested a mechanistic link between Notch signaling and resistance to apoptosis in melanoma [76]. Inhibition of basal Notch processing with gamma secretase inhibitors (GSIs) was observed to promote apoptosis in melanoma cell lines. This was associated with upregulated expression of the pro-apoptotic protein NOXA [77]. These results suggest that GSIs, which are currently being explored as a cancer therapeutic in T-lineage acute lymphoblastic leukemia [78], are attractive candidates to be considered to promote apoptosis in melanoma. For consideration of integration with anti-melanoma immunotherapy, careful analysis of the effects of GSIs on $CD8^+$ T cell activation, differentiation, and effector function will be required.

Opportunities for Therapeutic Intervention

As the development of antigen-specific immunotherapy for melanoma becomes further advanced, integration of approaches to increase the sensitivity of tumor cells to apoptosis will likely be critical. A similar consideration applies to improving responsiveness to conventional chemotherapy. An antisense approach target the anti-apoptotic Bcl-2 was the first strategy to enter the clinical arena. A randomized clinical trial comparing chemotherapy with dacarbazine versus dacarbazine plus antisense Bcl-2 showed a prolongation of progression free survival and response rate with the combination. However, the primary endpoint of overall survival was not significantly different between the groups [79]. Future considerations for clinical translation include antisense constructs with combined effects on multiple Bcl-2 family members [56], small molecule inhibitors of Bcl-2 function [79], cell-permeable survivin antagonists [80], GSIs to block Notch processing [81], and agonistic TRAIL receptor antibodies [82]. One important aspect of future clinical studies will be a combined understanding of effects of these agents on tumor cells versus anti-tumor T cells, which also are tightly regulated through programmed cell death.

Conclusions

The acquired resistance of melanoma cells to CTL-induced apoptosis may represent an important final barrier to immune-mediated destruction. Tumor cells can lose sensitivity to apoptosis via death receptors, and can upregulate expression of serpins and resist granule-mediated killing. They also can upregulate expression of general anti-apoptotic proteins, such as Bcl-2 family members and survivin. Finally, Notch signaling in melanoma likely contributes to an anti-apoptotic program. Development of clinically useful agents to counter these anti-apoptotic mechanisms in a tumor-targeted fashion should receive a high priority in the future.

References

[1] Dunn, G. P., Bruce, A. T., Ikeda, H., Old, L. J. & Schreiber, R. D. (2002) Cancer immunoediting: from immunosurveillance to tumor escape, *Nat Immunol. 3*, 991-8.

[2] Rosenberg, S. A. (2001) Progress in human tumour immunology and immunotherapy, *Nature. 411*, 380-4.

[3] Gajewski, T. F., Meng, Y. & Harlin, H. (2006) Immune suppression in the tumor microenvironment, *J Immunother. 29*, 233-40.

[4] Schwartz, R. H. (2003) T cell anergy, *Annu Rev Immunol. 21*, 305-34.

[5] Appleman, L. J. & Boussiotis, V. A. (2003) T cell anergy and costimulation, *Immunol Rev. 192*, 161-80.

[6] Chattopadhyay, S., Chakraborty, N. G. & Mukherji, B. (2005) Regulatory T cells and tumor immunity, *Cancer Immunol Immunother. 54*, 1153-61.

[7] Bluestone, J. A. & Boehmer, H. (2006) Regulatory T cells, *Semin Immunol. 18*, 77.

[8] Harlin, H., Kuna, T. V., Peterson, A. C., Meng, Y. & Gajewski, T. F. (2006) Tumor progression despite massive influx of activated CD8(+) T cells in a patient with malignant melanoma ascites, *Cancer Immunol Immunother*, 1-13.

[9] Leibson, P. J. (2004) The regulation of lymphocyte activation by inhibitory receptors, *Curr Opin Immunol. 16*, 328-36.

[10] Parry, R. V., Chemnitz, J. M., Frauwirth, K. A., Lanfranco, A. R., Braunstein, I., Kobayashi, S. V., Linsley, P. S., Thompson, C. B. & Riley, J. L. (2005) CTLA-4 and PD-1 receptors inhibit T-cell activation by distinct mechanisms, *Mol Cell Biol. 25*, 9543-53.

[11] Zha, Y., Blank, C. & Gajewski, T. F. (2004) Negative regulation of T-cell function by PD-1, *Crit Rev Immunol. 24*, 229-37.

[12] Blank, C., Gajewski, T. F. & Mackensen, A. (2005) Interaction of PD-L1 on tumor cells with PD-1 on tumor-specific T cells as a mechanism of immune evasion: implications for tumor immunotherapy, *Cancer Immunol Immunother. 54*, 307-14.

[13] Uyttenhove, C., Pilotte, L., Theate, I., Stroobant, V., Colau, D., Parmentier, N., Boon, T. & Van den Eynde, B. J. (2003) Evidence for a tumoral immune resistance mechanism based on tryptophan degradation by indoleamine 2,3-dioxygenase, *Nat Med. 9*, 1269-74.

[14] Russell, J. H. & Ley, T. J. (2002) Lymphocyte-mediated cytotoxicity, *Annu Rev Immunol. 20*, 323-70.

[15] Trapani, J. A. & Smyth, M. J. (2002) Functional significance of the perforin/granzyme cell death pathway, *Nat Rev Immunol. 2*, 735-47.

[16] Trapani, J. A. & Sutton, V. R. (2003) Granzyme B: pro-apoptotic, antiviral and antitumor functions, *Curr Opin Immunol. 15*, 533-43.

[17] Froelich, C. J., Metkar, S. S. & Raja, S. M. (2004) Granzyme B-mediated apoptosis--the elephant and the blind men?, *Cell Death Differ. 11*, 369-71.

[18] Thomas, D. A., Du, C., Xu, M., Wang, X. & Ley, T. J. (2000) DFF45/ICAD can be directly processed by granzyme B during the induction of apoptosis, *Immunity. 12*, 621-32.

[19] Pinkoski, M. J., Hobman, M., Heibein, J. A., Tomaselli, K., Li, F., Seth, P., Froelich, C. J. & Bleackley, R. C. (1998) Entry and trafficking of granzyme B in target cells during granzyme B-perforin-mediated apoptosis, *Blood. 92*, 1044-54.

[20] Metkar, S. S., Wang, B., Aguilar-Santelises, M., Raja, S. M., Uhlin-Hansen, L., Podack, E., Trapani, J. A. & Froelich, C. J. (2002) Cytotoxic cell granule-mediated apoptosis: perforin delivers granzyme B-serglycin complexes into target cells without plasma membrane pore formation, *Immunity. 16*, 417-28.

[21] Motyka, B., Korbutt, G., Pinkoski, M. J., Heibein, J. A., Caputo, A., Hobman, M., Barry, M., Shostak, I., Sawchuk, T., Holmes, C. F., Gauldie, J. & Bleackley, R. C. (2000) Mannose 6-phosphate/insulin-like growth factor II receptor is a death receptor for granzyme B during cytotoxic T cell-induced apoptosis, *Cell. 103*, 491-500.

[22] Millikin, D., Meese, E., Vogelstein, B., Witkowski, C. & Trent, J. (1991) Loss of heterozygosity for loci on the long arm of chromosome 6 in human malignant melanoma, *Cancer Res. 51*, 5449-53.

[23] Sun, J., Bird, C. H., Sutton, V., McDonald, L., Coughlin, P. B., De Jong, T. A., Trapani, J. A. & Bird, P. I. (1996) A cytosolic granzyme B inhibitor related to the viral apoptotic regulator cytokine response modifier A is present in cytotoxic lymphocytes, *J Biol Chem. 271*, 27802-9.

[24] Bird, C. H., Sutton, V. R., Sun, J., Hirst, C. E., Novak, A., Kumar, S., Trapani, J. A. & Bird, P. I. (1998) Selective regulation of apoptosis: the cytotoxic lymphocyte serpin proteinase inhibitor 9 protects against granzyme B-mediated apoptosis without perturbing the Fas cell death pathway, *Mol Cell Biol. 18*, 6387-98.

[25] Medema, J. P., de Jong, J., Peltenburg, L. T., Verdegaal, E. M., Gorter, A., Bres, S. A., Franken, K. L., Hahne, M., Albar, J. P., Melief, C. J. & Offringa, R. (2001) Blockade of the granzyme B/perforin pathway through overexpression of the serine protease inhibitor PI-9/SPI-6 constitutes a mechanism for immune escape by tumors, *Proc Natl Acad Sci USA. 98*, 11515-20.

[26] Phillips, T., Opferman, J. T., Shah, R., Liu, N., Froelich, C. J. & Ashton-Rickardt, P. G. (2004) A role for the granzyme B inhibitor serine protease inhibitor 6 in CD8+ memory cell homeostasis, *J Immunol. 173*, 3801-9.

[27] Zhang, M., Park, S. M., Wang, Y., Shah, R., Liu, N., Murmann, A. E., Wang, C. R., Peter, M. E. & Ashton-Rickardt, P. G. (2006) Serine protease inhibitor 6 protects

cytotoxic T cells from self-inflicted injury by ensuring the integrity of cytotoxic granules, *Immunity. 24*, 451-61.

[28] Li, C., Larson, D., Zhang, Z., Liu, Z., Strom, S. S., Gershenwald, J. E., Prieto, V. G., Lee, J. E., Ross, M. I., Mansfield, P. F., Cormier, J. N., Duvic, M., Grimm, E. A. & Wei, Q. (2006) Polymorphisms of the FAS and FAS ligand genes associated with risk of cutaneous malignant melanoma, *Pharmacogenet Genomics. 16*, 253-63.

[29] Mouawad, R., Khayat, D. & Soubrane, C. (2000) Plasma Fas ligand, an inducer of apoptosis, and plasma soluble Fas, an inhibitor of apoptosis, in advanced melanoma, *Melanoma Res. 10*, 461-7.

[30] Midis, G. P., Shen, Y. & Owen-Schaub, L. B. (1996) Elevated soluble Fas (sFas) levels in nonhematopoietic human malignancy, *Cancer Res. 56*, 3870-4.

[31] Ugurel, S., Rappl, G., Tilgen, W. & Reinhold, U. (2001) Increased soluble CD95 (sFas/CD95) serum level correlates with poor prognosis in melanoma patients, *Clin Cancer Res. 7*, 1282-6.

[32] Cheng, J., Zhou, T., Liu, C., Shapiro, J. P., Brauer, M. J., Kiefer, M. C., Barr, P. J. & Mountz, J. D. (1994) Protection from Fas-mediated apoptosis by a soluble form of the Fas molecule, *Science. 263*, 1759-62.

[33] Mendez, R., Ruiz-Cabello, F., Rodriguez, T., Del Campo, A., Paschen, A., Schadendorf, D. & Garrido, F. (2006) Identification of different tumor escape mechanisms in several metastases from a melanoma patient undergoing immunotherapy, *Cancer Immunol Immunother.*

[34] Meng, Y., Kang, S., So, J., Reierstad, S. & Fishman, D. A. (2005) Translocation of Fas by LPA prevents ovarian cancer cells from anti-Fas-induced apoptosis, *Gynecol Oncol. 96*, 462-9.

[35] Algeciras-Schimnich, A., Shen, L., Barnhart, B. C., Murmann, A. E., Burkhardt, J. K. & Peter, M. E. (2002) Molecular ordering of the initial signaling events of CD95, *Mol Cell Biol. 22*, 207-20.

[36] Barnhart, B. C., Legembre, P., Pietras, E., Bubici, C., Franzoso, G. & Peter, M. E. (2004) CD95 ligand induces motility and invasiveness of apoptosis-resistant tumor cells, *Embo J. 23*, 3175-85.

[37] Whiteside, T. L. (2002) Tumor-induced death of immune cells: its mechanisms and consequences, *Semin Cancer Biol. 12*, 43-50.

[38] Tschopp, J., Irmler, M. & Thome, M. (1998) Inhibition of fas death signals by FLIPs, *Curr Opin Immunol. 10*, 552-8.

[39] French, L. E. & Tschopp, J. (2002) Defective death receptor signaling as a cause of tumor immune escape, *Semin Cancer Biol. 12*, 51-5.

[40] Song, J. H., Song, D. K., Herlyn, M., Petruk, K. C. & Hao, C. (2003) Cisplatin down-regulation of cellular Fas-associated death domain-like interleukin-1beta-converting enzyme-like inhibitory proteins to restore tumor necrosis factor-related apoptosis-inducing ligand-induced apoptosis in human melanoma cells, *Clin Cancer Res. 9*, 4255-66.

[41] Irmler, M., Thome, M., Hahne, M., Schneider, P., Hofmann, K., Steiner, V., Bodmer, J. L., Schroter, M., Burns, K., Mattmann, C., Rimoldi, D., French, L. E. & Tschopp, J. (1997) Inhibition of death receptor signals by cellular FLIP, *Nature. 388*, 190-5.

[42] Pitti, R. M., Marsters, S. A., Ruppert, S., Donahue, C. J., Moore, A. & Ashkenazi, A. (1996) Induction of apoptosis by Apo-2 ligand, a new member of the tumor necrosis factor cytokine family, *J Biol Chem. 271*, 12687-90.

[43] Pan, G., O'Rourke, K., Chinnaiyan, A. M., Gentz, R., Ebner, R., Ni, J. & Dixit, V. M. (1997) The receptor for the cytotoxic ligand TRAIL, *Science. 276*, 111-3.

[44] Walczak, H., Degli-Esposti, M. A., Johnson, R. S., Smolak, P. J., Waugh, J. Y., Boiani, N., Timour, M. S., Gerhart, M. J., Schooley, K. A., Smith, C. A., Goodwin, R. G. & Rauch, C. T. (1997) TRAIL-R2: a novel apoptosis-mediating receptor for TRAIL, *Embo J. 16*, 5386-97.

[45] Zhang, X. D., Zhang, X. Y., Gray, C. P., Nguyen, T. & Hersey, P. (2001) Tumor necrosis factor-related apoptosis-inducing ligand-induced apoptosis of human melanoma is regulated by smac/DIABLO release from mitochondria, *Cancer Res. 61*, 7339-48.

[46] Zhang, X. D., Wu, J. J., Gillespie, S., Borrow, J. & Hersey, P. (2006) Human melanoma cells selected for resistance to apoptosis by prolonged exposure to tumor necrosis factor-related apoptosis-inducing ligand are more vulnerable to necrotic cell death induced by cisplatin, *Clin Cancer Res. 12*, 1355-64.

[47] Li, H., Niederkorn, J. Y., Neelam, S. & Alizadeh, H. (2005) Resistance and susceptibility of human uveal melanoma cells to TRAIL-induced apoptosis, *Arch Ophthalmol. 123*, 654-61.

[48] Hanahan, D. & Weinberg, R. A. (2000) The hallmarks of cancer, *Cell. 100*, 57-70.

[49] Garcia, I., Martinou, I., Tsujimoto, Y. & Martinou, J. C. (1992) Prevention of programmed cell death of sympathetic neurons by the bcl-2 proto-oncogene, *Science. 258*, 302-4.

[50] Tas, F., Duranyildiz, D., Argon, A., Oguz, H., Camlica, H., Yasasever, V. & Topuz, E. (2004) Serum bcl-2 and survivin levels in melanoma, *Melanoma Res. 14*, 543-6.

[51] Zhang, H. & Rosdahl, I. (2006) Bcl-xL and bcl-2 proteins in melanoma progression and UVB-induced apoptosis, *Int J Oncol. 28*, 661-6.

[52] Igney, F. H. & Krammer, P. H. (2002) Death and anti-death: tumour resistance to apoptosis, *Nat Rev Cancer. 2*, 277-88.

[53] Bush, J. A. & Li, G. (2003) The role of Bcl-2 family members in the progression of cutaneous melanoma, *Clin Exp Metastasis. 20*, 531-9.

[54] Selzer, E., Schlagbauer-Wadl, H., Okamoto, I., Pehamberger, H., Potter, R. & Jansen, B. (1998) Expression of Bcl-2 family members in human melanocytes, in melanoma metastases and in melanoma cell lines, *Melanoma Res. 8*, 197-203.

[55] Leiter, U., Schmid, R. M., Kaskel, P., Peter, R. U. & Krahn, G. (2000) Antiapoptotic bcl-2 and bcl-xL in advanced malignant melanoma, *Arch Dermatol Res. 292*, 225-32.

[56] Olie, R. A., Hafner, C., Kuttel, R., Sigrist, B., Willers, J., Dummer, R., Hall, J., Stahel, R. A. & Zangemeister-Wittke, U. (2002) Bcl-2 and bcl-xL antisense oligonucleotides induce apoptosis in melanoma cells of different clinical stages, *J Invest Dermatol. 118*, 505-12.

[57] Vucic, D., Stennicke, H. R., Pisabarro, M. T., Salvesen, G. S. & Dixit, V. M. (2000) ML-IAP, a novel inhibitor of apoptosis that is preferentially expressed in human melanomas, *Curr Biol. 10*, 1359-66.

[58] Schmollinger, J. C., Vonderheide, R. H., Hoar, K. M., Maecker, B., Schultze, J. L., Hodi, F. S., Soiffer, R. J., Jung, K., Kuroda, M. J., Letvin, N. L., Greenfield, E. A., Mihm, M., Kutok, J. L. & Dranoff, G. (2003) Melanoma inhibitor of apoptosis protein (ML-IAP) is a target for immune-mediated tumor destruction, *Proc Natl Acad Sci USA. 100*, 3398-403.

[59] Shankar, S. L., Mani, S., O'Guin, K. N., Kandimalla, E. R., Agrawal, S. & Shafit-Zagardo, B. (2001) Survivin inhibition induces human neural tumor cell death through caspase-independent and -dependent pathways, *J Neurochem. 79*, 426-36.

[60] Beltrami, E., Plescia, J., Wilkinson, J. C., Duckett, C. S. & Altieri, D. C. (2004) Acute ablation of survivin uncovers p53-dependent mitotic checkpoint functions and control of mitochondrial apoptosis, *J Biol Chem. 279*, 2077-84.

[61] Liu, T., Brouha, B. & Grossman, D. (2004) Rapid induction of mitochondrial events and caspase-independent apoptosis in Survivin-targeted melanoma cells, *Oncogene. 23*, 39-48.

[62] Li, F., Ambrosini, G., Chu, E. Y., Plescia, J., Tognin, S., Marchisio, P. C. & Altieri, D. C. (1998) Control of apoptosis and mitotic spindle checkpoint by survivin, *Nature. 396*, 580-4.

[63] Vong, Q. P., Cao, K., Li, H. Y., Iglesias, P. A. & Zheng, Y. (2005) Chromosome alignment and segregation regulated by ubiquitination of survivin, *Science. 310*, 1499-504.

[64] Fortugno, P., Beltrami, E., Plescia, J., Fontana, J., Pradhan, D., Marchisio, P. C., Sessa, W. C. & Altieri, D. C. (2003) Regulation of survivin function by Hsp90, *Proc Natl Acad Sci USA. 100*, 13791-6.

[65] Ambrosini, G., Adida, C. & Altieri, D. C. (1997) A novel anti-apoptosis gene, survivin, expressed in cancer and lymphoma, *Nat Med. 3*, 917-21.

[66] Velculescu, V. E., Madden, S. L., Zhang, L., Lash, A. E., Yu, J., Rago, C., Lal, A., Wang, C. J., Beaudry, G. A., Ciriello, K. M., Cook, B. P., Dufault, M. R., Ferguson, A. T., Gao, Y., He, T. C., Hermeking, H., Hiraldo, S. K., Hwang, P. M., Lopez, M. A., Luderer, H. F., Mathews, B., Petroziello, J. M., Polyak, K., Zawel, L., Kinzler, K. W. & et al. (1999) Analysis of human transcriptomes, *Nat Genet. 23*, 387-8.

[67] Takeuchi, H., Morton, D. L., Elashoff, D. & Hoon, D. S. (2005) Survivin expression by metastatic melanoma predicts poor disease outcome in patients receiving adjuvant polyvalent vaccine, *Int J Cancer. 117*, 1032-8.

[68] Chawla-Sarkar, M., Bae, S. I., Reu, F. J., Jacobs, B. S., Lindner, D. J. & Borden, E. C. (2004) Downregulation of Bcl-2, FLIP or IAPs (XIAP and survivin) by siRNAs sensitizes resistant melanoma cells to Apo2L/TRAIL-induced apoptosis, *Cell Death Differ. 11*, 915-23.

[69] Grossman, D., Kim, P. J., Schechner, J. S. & Altieri, D. C. (2001) Inhibition of melanoma tumor growth in vivo by survivin targeting, *Proc Natl Acad Sci USA. 98*, 635-40.

[70] Mesri, M., Wall, N. R., Li, J., Kim, R. W. & Altieri, D. C. (2001) Cancer gene therapy using a survivin mutant adenovirus, *J Clin Invest. 108*, 981-90.

[71] Hansson, E. M., Lendahl, U. & Chapman, G. (2004) Notch signaling in development and disease, *Semin Cancer Biol. 14*, 320-8.

[72] Wu, L. & Griffin, J. D. (2004) Modulation of Notch signaling by mastermind-like (MAML) transcriptional co-activators and their involvement in tumorigenesis, *Semin Cancer Biol. 14*, 348-56.

[73] Larsson, C., Lardelli, M., White, I. & Lendahl, U. (1994) The human NOTCH1, 2, and 3 genes are located at chromosome positions 9q34, 1p13-p11, and 19p13.2-p13.1 in regions of neoplasia-associated translocation, *Genomics. 24*, 253-8.

[74] Lefort, K. & Dotto, G. P. (2004) Notch signaling in the integrated control of keratinocyte growth/differentiation and tumor suppression, *Semin Cancer Biol. 14*, 374-86.

[75] Collins, B. J., Kleeberger, W. & Ball, D. W. (2004) Notch in lung development and lung cancer, *Semin Cancer Biol. 14*, 357-64.

[76] Nickoloff, B. J., Osborne, B. A. & Miele, L. (2003) Notch signaling as a therapeutic target in cancer: a new approach to the development of cell fate modifying agents, *Oncogene. 22*, 6598-608.

[77] Qin, J. Z., Ziffra, J., Stennett, L., Bodner, B., Bonish, B. K., Chaturvedi, V., Bennett, F., Pollock, P. M., Trent, J. M., Hendrix, M. J., Rizzo, P., Miele, L. & Nickoloff, B. J. (2005) Proteasome inhibitors trigger NOXA-mediated apoptosis in melanoma and myeloma cells, *Cancer Res. 65*, 6282-93.

[78] Grabher, C., von Boehmer, H. & Look, A. T. (2006) Notch 1 activation in the molecular pathogenesis of T-cell acute lymphoblastic leukaemia, *Nat Rev Cancer. 6*, 347-59.

[79] Hersey, P. (2006) Apoptosis and melanoma: how new insights are effecting the development of new therapies for melanoma, *Curr Opin Oncol. 18*, 189-96.

[80] Yan, H., Thomas, J., Liu, T., Raj, D., London, N., Tandeski, T., Leachman, S. A., Lee, R. M. & Grossman, D. (2006) Induction of melanoma cell apoptosis and inhibition of tumor growth using a cell-permeable Survivin antagonist, *Oncogene.*

[81] Curry, C. L., Reed, L. L., Golde, T. E., Miele, L., Nickoloff, B. J. & Foreman, K. E. (2005) Gamma secretase inhibitor blocks Notch activation and induces apoptosis in Kaposi's sarcoma tumor cells, *Oncogene. 24*, 6333-44.

[82] Buchsbaum, D. J., Zhou, T. & Lobuglio, A. F. (2006) TRAIL receptor-targeted therapy, *Future Oncol. 2*, 493-508.

In: Cell Apoptosis and Cancer
Editor: Albina W. Taylor, pp. 87-111
ISBN: 978-1-60021-506-3

Chapter VI

Apoptosis of Leukemia Cells Induced by L-Ascorbic Acid and Arsenic Trioxide: The Effect of Oxidative Stress and Glutathione Homeostasis

Seyeon Park
Department of Applied Chemistry, Dongduk Women's University,
Seoul 136-714, Korea.

Abstract

L-ascorbic acid (LAA) at concentrations of 0.25–1.0mM induced a dose- and time-dependent inhibition of proliferation in acute myeloid leukemia (AML) cell lines and also in leukemic cells from peripheral blood specimens obtained from three patients with AML. LAA induced oxidation of glutathione to oxidized form (GSSG) and subsequent H_2O_2 accumulation in a concentration-dependent manner, in parallel to induction of apoptosis. Induction of apoptosis in LAA-treated AML cells involved a dose-dependent increase of Bax protein, release of cytochrome C from mitochondria to cytosol, activation of caspase 9 and caspase 3, and cleavage of poly [ADP-ribose] polymerase. LAA treatment of cells induces a dose-dependent phosphorylation of extracellular signalregulated kinases (ERK) and results in activation of its catalytic domain. The small G protein Raf1 and MAPKactivated protein kinase 2 are activated by LAA as an upstream and a downstream regulator of ERK, respectively. Although the ERK pathway has been known to activate cell proliferation, pharmacologic inhibition of ERK reduces LAA-dependent apoptosis and growth inhibitory response of acute myeloid leukemia cell lines, suggesting that this signaling cascade positively regulates induction of apoptotic response by LAA.

Arsenic trioxide (As_2O_3) could induce *in vitro* growth inhibition and apoptosis of AML cells. As_2O_3 or LAA suppressed DNA-binding activity of NF-κB composed of p65/p50 heterodimer through preventing the degradation of IκB-α and the nuclear translocation of p65 subsequently as well as interrupting the binding of NF-κB with their consensus sequences. Inhibitory effect of As_2O_3 or LAA on NF-κB DNA activity was dependent upon intracellular glutathione (GSH) and H_2O_2 level, but not superoxide anion.

The addition of LAA to cells resulted in an immediate increase in intracellular total GSH content and GST activity accompanied by the uptake of cysteine. These results suggest a new role of LAA in intracellular sulfur condition and glutathione regulation. Early protein targets of LAA in leukemia cells were identified using a differential proteomics approach in NB4 human leukemia cells treated with LAA. A subunit of protein-disulfide isomerase (a thiol/disulfide exchange catalyst) and immunoglobulin-heavy-chain binding protein (BiP, identical to Hsp70 chaperone) showed quantitative expression profile differences. A myeloid leukemia associated antigen protein (a tropomyosin isoform) showed changes in pI as a result of phosphorylation. The addition of LAA to cells results in an immediate change in the intracellular thiol/disulfide condition and modulates disulfide bond formation in proteins.

INTRODUCTION

Apoptosis has emerged as an important biologic mechanism that contributes to the maintenance of the integrity of multicellular organisms. Impairment of apoptosis has been implicated in any human diseases including cancers [1-3]. For instance, deregulation of apoptosis is regarded as a major factor contributing to the development of chronic lymphocytic leukemia (CLL) and some types of lymphoma, such as follicular B-cell lymphoma with the chromosomal translocation t(14:18), which results in overexpression of the death antagonist Bcl-2 [4-6]. Therefore, understanding the basic mechanisms that underlie apoptosis will help identify new potential targets for treatment of these diseases [1,7,8].

Recently, arsenic trioxide (As_2O_3), an ancient drug used in Traditional Chinese Medicine, attracted wide interest for its ability to induce complete remission in most patients with acute promyelocytic leukemia (APL), mainly through the induction of apoptosis and differentiation [1,9-14]. Moreover, it was reported that clinically achievable concentrations of As_2O_3 can also trigger apoptosis in chronic myelogenous leukemia cells, some multiple myeloma cells, and some solid tumor cells, such as esophageal cancer cancer and neuroblastoma cells, suggesting that the ability of As_2O_3 to induce apoptosis is not limited to APL [1]. There are data suggesting that oxidative stress triggers signal transduction cascade leading to apoptosis by arsenic trioxide [15-18].

Meanwhile, LAA in combination with arsenic trioxide has shown clinical responses in chemorefractory multiple meyloma [15]. There is also increasing evidence that LAA is selectively toxic to some types of tumor cells, functioning as a pro-oxidant, rather than anti-oxidant [19-21]. LAA at concentrations of 10 nM–1mM induced apoptosis in neuroblastoma and melanoma cells [22]. LAA was also shown to be an important modulator for the growth of mouse myeloma cells in an in vitro colony assay [23]. A series of studies have established that the growth of leukemic progenitor cells from patients with acute myeloid leukemia (AML) and myelodysplastic syndromes (MDS) can be profoundly modulated by LAA[24,25]. Intravenous administration of sodium 5,6-benzylidene-l-ascorbate (SBA) to inoperable cancer patients induced a significant reduction in the tumor volume, without any adverse side effects [26]. Further, recent clinical studies indicate that manipulation of LAA levels in vivo can result in clinical benefit for patients with AML and solid tumors [27].

In this chapter, I summarized the common characteristics and differences of the apoptois mechanisms of leukemia cells induced by As_2O_3 and LAA, highlighting the oxidative stress.

Apoptosis and Growth Inhibition in Leukemia Cells after Treatment with L-Ascorbic Acid or Arsenic Trioxide

Treatment of malignant lymphocytic cell lines with LAA (0.25–1 mM) for 24 h led to a marked dose-dependent decrease of cell proliferation [28].The responsive cell lines were human myeloid leukemia cell line HL-60, retinoic acid (RA)-sensitive acute promyelocytic leukemia (APL) cell line NB4 and RA-resistant APL cell line NB4-R1. Different types of leukemia cells such as K562 (chronic myelogenous leukemia cell line) [28] and KG1 (myeloblast cell line) (our unpublished data) were also responsive to LAA. A similar result was obtained with cells containing over 90% blasts from patients with AML (Table 1). In contrast, exposure of three ovarian cancer cell lines including SK-OV-3, OVCAR-3 and 2774 to the same concentrations of LAA had little effect on the proliferation of these cells [28]. The differential sensitivity could be partly associated with differences in LAA uptake when compared to AML cell lines, showing lower uptake efficiency in ovarian cancer cell lines [28].

Our previous results indicated that for three acute myeloid leukemia cell lines (HL-60, NB4 and NB4-R1), 0.5 mM LAA for 24 h induced apoptosis in approximately 16–26% of the treated cells [28]. Induction of apoptosis by LAA in these cell lines demonstrated a dose-dependent pattern. LAA failed to promote apoptosis in ovarian cell lines including SK-OV-3, OVCAR-3 and 2774 [28]. As compared with control cells, LAA treatment resulted in increases in the G1 fraction ranging from 7% (0.25 mM) to 25% (1 mM), with concomitant reduction in the S phase fraction of 3 and 30%, respectively. This indicates the induction of a G1/S arrest by LAA.

As_2O_3 also exerted substantial dose- and time-dependent growth inhibition in malignant lymphocytic cell lines, but different cells exhibited distinct sensitivities to As_2O_3 [1]. In the previous study, the responses of several malignant lymphocytic cell lines and primary acute lymphocytic leukemia (ALL), chronic lymphocytic leukemia (CLL), and lymphoma cells to a wide range of As_2O_3 concentrations (0.1–2 μM) were investigated. As_2O_3 at 0.1–0.25 μM did not substantially inhibit the growth of most cells during 3–5 days of treatment [1]. At 1–2 μM As_2O_3, however, growth inhibition was evident after 1–3 days and became substantial after 3–5 days in almost all cell lines (Table 2). In our experiment, As_2O_3 caused the growth inhibition of HL-60 cells stronger than normal bone marrow (BM) cells [29]. Such a growth inhibitory effect was accompanied by cell apoptosis. The decrease in cell viability was accompanied by the appearance of morphologic characteristics of apoptosis, such as shrinking cytoplasm, condensed chromatin, and nuclear fragmentation with intact cell membrane. Cultures of these cells demonstrated a dose- and time-dependent increase in the number of cells with a sub-G1 DNA content [29].

Although 1-2 μ*M* As_2O_3 inhibited the growth of nearly all cell lines examined (table 2), it did not induce apoptosis in Nalm-6, Raji, and Jurkat cells. It appears to exert a cytostatic effect in these cells [1]. Zhu and coworkers observed a dose- and time-dependent increase in the number of cells with a sub-G1 DNA content in SKW-3 and Molt-4 cells [1]. Also they observed DNA fragmentation in 1 m*M* As_2O_3-treated SKW-3 cells and 2 m*M* As_2O_3-treated Namalwa cells. 2 μ*M* As_2O_3 induced apoptosis in HL-60 [29] NB4 and NB4- R1 (our unpublished data). These results suggested that 2 μ*M* As_2O_3 or less induced apoptosis in most malignant lymphocytic cell lines studied.

The above data suggest that LAA and As_2O_3 treatment had dual effects on malignant lymphocytic cells, i.e., growth arrest and apoptosis.

Table 1. Effects of LAA treatment on cell growth and apoptosis.

LAA	Cell line	% Viability	Apoptotic cell %
(250 μM, 500 μM)	HL-60	70.0 ± 3.0	6.8 ± 3.2
		29.6 ± 8.8	24.6 ± 11.0
	NB4	100 ± 11.4	3.1 ± 2.1
		79.4 ± 3.5	15.9 ± 9.0
	NB4-R1	100 ± 4.2	4.4 ± 2.0
		40.5 ± 9.3	19.1 ± 6.9
	NB4-R2	100 ± 3.0	2.6 ± 3.2
		79.1 ± 12.8	21.0 ± 10.1
	KG1	99.6 ± 2.2	5.1 ± 7.8
		78.5 ± 8.4	18.3 ± 3.6
	K562	80.1 ± 1.1	6.3 ± 2.7
		49.5 ± 6.8	21.9 ± 2.9
	U937	61.9 ± 3.6	
		19.5 ± 4.2	
	Normal BM	88 ± 11.7	
		75 ± 7.8	
	Patient with	84.8 ± 10.6	
	AML	69.1 ± 9.6	
	OVCAR	94.8 ± 1.2	6.2 ± 2.1
		91.9 ± 2.3	6.3 ± 3.8
	SK-OV3	100 ± 8.4	0.1 ± 1.6
		98 ± 6.0	0.1 ± 3.0

Data are means ± standard deviations of triplicate experiments. HL-60, human myeloid leukemia; NB4, NB4-R1, NB4-R2, human acute promyelocytic leukemia (APL); KG1 human myeloblast; K562 human chronic myelogenous leukemia; U937, human histiocytic lymphoma; OVCAR, SK-OV3, ovarian cancer.

Table 2. Effects of As_2O_3 treatment on cell growth and apoptosis.

As_2O_3(1 μM)	*Cell line*	*% Viability*	*Apoptotic cell %*
1day	HL-60	93.4 ± 3.6 [29]	7.2 (control 4.6%)[29]
3day		84.8 [15]	9.0 ± 1.7 (control 4.0%)[15]
	NB4		33 [15] *
		25 [15] *	57.8 ± 7.2 [18]
	KG1		17.5 ± 4.2(48 h)[18]
	K562		6.7 ± 3.1(48 h)[18]
	U937		5.6 ± 3.4 (48 h)[18]
	SKW-3 [1]	87.0 ± 2.9	10 [1] *
		75.3 ± 1.3	25 [1] *
	Molt-4 [1]	96.4 ± 5.1	
		60.6 ± 3.4	
	BJAB [1]	94.0 ± 2.6	
		85.4 ± 1.1	
	Su-DHL-4 [1]	92.0 ± 2.7	10.5 [15] *
		82.0 ± 3.1	12.5 [15] *
	Namalwa [1]	97.4 ± 3.2	3 [1] *
		90.6 ± 4.5	4 [1] *
	Nalm-6 [1]		
		97.7 ± 2.5	
	Jurkat [1]	95 .8 ± 0.6	
		97.0 ± 1.8	
	Raji [1]	93.2 ± 2.0	
		88.5 ± 3.3	

Data are means ± standard deviations of triplicate experiments. HL-60, human myeloid leukemia; NB4, human acute promyelocytic leukemia (APL); KG1 human myeloblast; K562 human chronic myelogenous leukemia; U937, human histiocytic lymphoma; SKW-3, human T-cell chronic lymphocytic leukemia; Molt-4 and Jurkat, human T-cell acute lymphocytic leukemia; BJAB, B cell lymphoma; su-DHL-4, follicular B-cell lymphoma; Namalwa and Raji, Burkitt's lymphoma; Nalm-6, pre-B-cell acute lymphocytic leukemia.

* Data are approximate values determined from the graph.

Correlation of Apoptotic Effect with Redox Status Modulation

It is well known that an important function of GSH is the elimination of reactive oxygen species (ROS) and the reduction of dehydroascorbic acid (DHA) to LAA by GSH-dependent mechanisms in the cell [30]. The correlation of the GSH level with the inhibitory effect of LAA on the proliferation of HL-60, NB4 cells and cells from a patient with AML was investigated [28]. These AML cells were cultured with various concentrations of LAA in the presence or absence of buthione sulfoximine (BSO) or dithiothreitol (DTT) for 24 h, respectively. The addition of 1mM BSO, which inhibited GSH synthesis resulting in lower GSH levels, increased the growth inhibition of HL-60 cells induced by LAA. However, the treatment of 0.2mM DTT, which reduced GSSG back to GSH, decreased the LAA-induced growth inhibition of HL-60 cells [28]. A striking finding is that the inhibitory effect of LAA was completely abrogated by the addition of 500 U/ml catalase, regardless of LAA concentration; whereas the addition of 100 μM NaN_3 synergistically enhanced the effect of LAA [28]. These results were also observed in NB4 cells and in cells from a patient with AML [28]. These results indicated that the removal of H_2O_2 by catalase completely abolished the inhibitory effect of LAA; but the inhibition of endogenous catalase by NaN_3 accelerated it. The inhibitory effect of LAA was not significantly affected by the addition of 1000 U/ml superoxide dismutase (SOD). This result indicated that superoxide anion had no appreciable role in the inhibitory effect of LAA. The results indicated that at least the major portion, if not all, of the growth inhibitory effect of LAA is due to the generation of H_2O_2.

In accordance with cell proliferation assay, LAA-induced apoptosis was affected by the modulation of cellular redox status [28]. The addition of BSO increased the apoptosis in HL-60 cells induced by LAA. However, treatment of DTT decreased the LAA-induced apoptosis of HL-60 cells. On the other hand, treatment with 1 mM BSO alone for 24 hours rarely induced apoptosis in these cells. The results reported that the removal of H_2O_2 by catalase completely abolished the induction of apoptosis by LAA.

Zu and co-workers treated SKW-3, Namalwa, and Jurkat cells simultaneously with 1–2 mM As_2O_3 and 0.2 mM DTT or 1 mM BSO. DTT treatment alone had no substantial effect on mitochondrial membrane potential ($\Delta\psi_m$), cell viability, and the percentage of sub-G1 cells in SKW-3 and Namalwa cells. However, in these cell lines, 0.2 mM DTT blocked the $\Delta\psi_m$ collapse induced by 2 mM. As_2O_3 DTT also blocked the As_2O_3-induced decrease in cell viability, ATP depletion, and the increase in the sub-G1 cell population.

Although 1 mM As_2O_3 had no effect on cell viability and $\Delta\psi_m$ in Namalwa cells, simultaneous treatment with 1 mM As_2O_3 and 1 mM BSO substantially enhanced the magnitude of $\Delta\psi_m$ collapse and apoptosis. This cooperative effect was also evident in Jurkat cells [1] and HL-60 cells [29].

EFFECTS OF LAA AND ARSENIC TRIOXIDE ON INTRACELLULAR GSH OXIDATION STATE

The glutathione (GSH) redox system is known to modulate the growth-inhibitory effect of arsenicals [15,31-36]. As^{5+}, cadmium, thalium, selenium, zinc, or mercury did not induce apoptosis at 1 μM in NB4 cells, suggesting an effect of As^{3+} (as in As_2O_3) in this process [14]. The findings that arsenites (40 μM) can induce apoptosis in CHO cells [37], that hydrogen peroxide-resistant CHO cells are less responsive, and that catalase-deficient CHO cells are hypersensitive suggest an important role for hydrogen peroxide as a mediator of arsenic-induced apoptosis [38]. Recently, however, trivalent antimonial compounds (Sb-III), which belongs to the same group of the periodic table (Group V), were investigated as effective apoptosis-triggering compounds [39]. They suggested that antimony induced GSH efflux from THP-macrophages and apoptosis in these cells.

The growth inhibitory and apoptotic effect of As_2O_3 inversely correlates with GSH content of NB4, su-DHL-4, and HL-60 cells [15]. According to the report [15], the GSH content was three times greater in HL-60 cells than in NB4 cells, whereas in su-DHL-4 cells GSH content fell in between these two values. Moreover, Cell growth inhibition and apoptotic effect of As_2O_3 is modulated by experimentally changing the GSH content. Increasing the GSH content by N-acetylcysteine (NAC) treatment [40] in NB4 cells prevented 2 μM As_2O_3 induction of apoptosis [15]. Similarly, lipoic acid completely blocked As_2O_3-induced apoptosis, perhaps by binding As_2O_3 to its vicinal thiol groups and increasing GSH content [41]. Both NAC and lipoic acid treatment inhibited As_2O_3-induced activation of CPP32 and apoptosis. However, NAC did not and lipoic acid only partially inhibited the degradation of PML/RAR-a protein. Thus, the degradation of PML/RAR-a alone does not appear to be sufficient for the As_2O_3 induction of apoptosis.

Dai et al reported that ascorbic acid enhanced As_2O_3 antilymphoma effect from the study using mouse lymphoma P388D1 cells, which demonstrated in vitro ascorbic acid enhancement of As_2O_3 growth inhibition and were implanted into the peritoneum of BDF1 mice [15]. The combined treatment of As_2O_3 (5 mg/kg) and ascorbic acid (500 mg/kg) improved survival time, whereas single-agent treatment at these nontoxic concentrations did not influence survival.

Several groups have shown that LAA augments As_2O_3-mediated apoptosis [15,17]. Our group, however, presented data that LAA alone could modulate malignant cell growth in vitro [28]. Also, LAA alone could effectively induce apoptosis of the AML cell lines HL-60, NB4 and NB4-R1. From our experiments, incubation with LAA for 1 h resulted in the decrease of GSH/GSSG ratio in not only three myeloid leukemia cell lines but also cells from patients with AML, in concentration dependent manners. Interestingly, however, incubation with LAA did not show such decrease of GSH/GSSG ratio in three ovarian cancer cell lines [28], suggesting that the differential sensitivity to LAA-induced apoptosis between cell lines was associated with differences in cellular oxidation states.

Effects of LAA on Intracellular GSH Homeostasis

From our experiments, LAA treatment changed the GSH oxidation state in cell lines such as HL-60, NB4, NB4-R1 [28] and KG1 (unpublished data). As shown in Figure 1A, incubation with LAA for 1 hr resulted in the decrease of GSH/GSSG ratio in a concentration-dependent manner. However, incubating AML cells with LAA resulted in a marked increase in the concentration of total glutathione relative to the total protein concentration, which reached a maximum after 2 or 3 hr of 163%, 200%, 197% for for HL-60 (Figure 1B), NB4, and KG1 cells, respectively, followed by a gradual decline. The stimulated responses to LAA were dependent on the concentration of LAA.

It follows that if LAA-mediated increase in intracellular GSH oxidation increased intracellular H_2O_2 level, the LAA is likely to induce intracellular oxidative stress via GSH/GSSG ratio changes. We tested the intracellular hydrogen peroxide accumulation physically using H_2DCFDA and flowcytometry analysis. Consistent with the proposed pro-oxidant activity, the LAA-mediated decrease in intracellular GSH/GSSG ratios correlated with intracellular increase in H_2O_2 level as shown (C). Additionally, LAA-mediated increase in the total GSH level after around 3 hr of treatment could impair the intracellular H_2O_2 level, explaining that the intracellular ratio of GSH and GSSG correlated functionally with a H_2O_2 accumulation, which results in cellular oxidative stress in HL-60 cells.

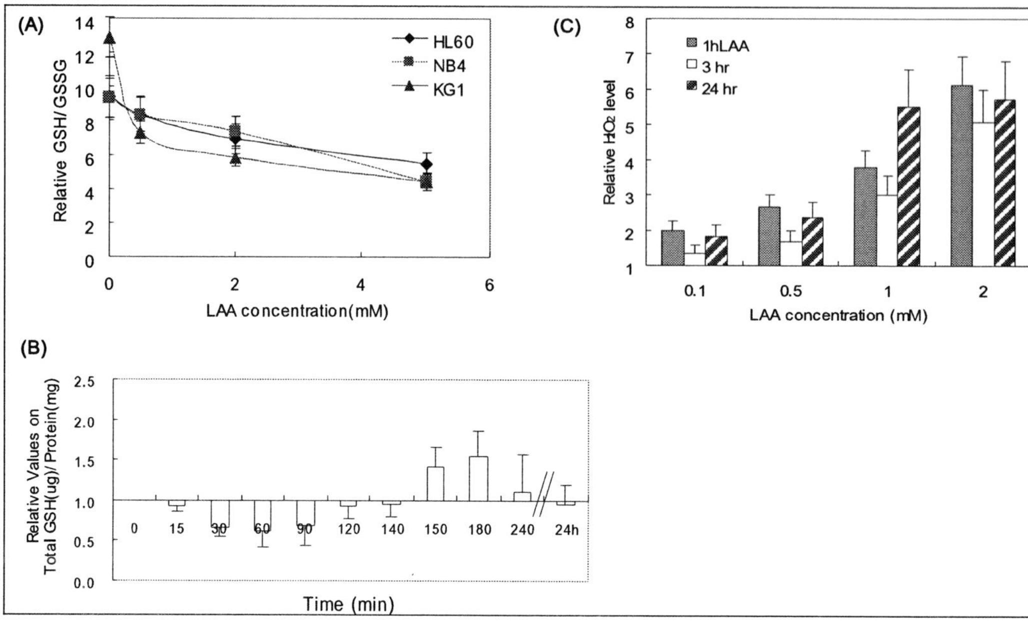

Figure1. The effect of LAA on the intracellular redox system. (A) The effect of LAA on the GSH/GSSG ratio. HL-60, KG1 and NB4 cells were incubated for 1 hr, and the intracellular GSH and GSSG level were estimated. Values are means ± standard deviations (S.D.) of four experiments. (B) Time course study on the effect of LAA on the concentration of total GSH. HL-60 cells were incubated with 1 mM of LAA, and changes in intracellular GSH were estimated. Values are means ± S.D. of four experiments (C) Analysis of H_2O_2 accumulation in LAA-treated HL-60 cells. HL-60 cells were labeled with H_2DCFDA fluorescent probe for 30 min and then treated with or without LAA for 1 hr, 3 hr and 24 hr. Fluorescence was analyzed by flowcytometry. Data are represented as relative values compared with control.

Effects of LAA on Intracellular -SH Uptake

It is generally believed that cysteine is a rate-limiting precursor for GSH synthesis [42]. We measured cysteine uptake in leukemia cell lines exposed to LAA. The intracellular L-Cys incorporation was measured in intact HL-60 and KG1 cells exposed to LAA by using ^{35}S-labeled-L-Cys in media. Figure 2 illustrates the time-dependent changes in the rate of the uptake of cysteine during culture. It was observed that the rate of uptake, determined under conditions without LAA was very low (at most 119% of baseline by 16 hr). However, it increased greatly showing two peaks after around 1 hr and 3 hr. A glutathione synthesis inhibitor BSO potently inhibited the second peak, suggesting that the incorporation of cysteine corresponding to the second peak is mediated by glutathione synthesis (unpublished data). Our results indicated that LAA induced glutathione synthesis accompanied by intracellular cysteine uptake. Glutathione-S transferases (GSTs) are a family of enzymes that catalyze the conjugation of electrophilic substitute to GSH, and play roles in protecting cells by removing reactive oxygen species and regenerating S-thiolated proteins. Interestingly, intracellular total GSH levels in these AML cells incubated with LAA peaked around 3 hr and declined (Figure 1B), while the increase in [^{35}S]-L-Cys incorporated conformed at 3 hr and kept on going (Figure. 2). This result demonstrated that [^{35}S]-L-Cys once transported into cells through cysteine uptake is incorporated and transferred intracellularly, causing us to hypothesize that the sulfhydryl transfer system is affected by LAA.

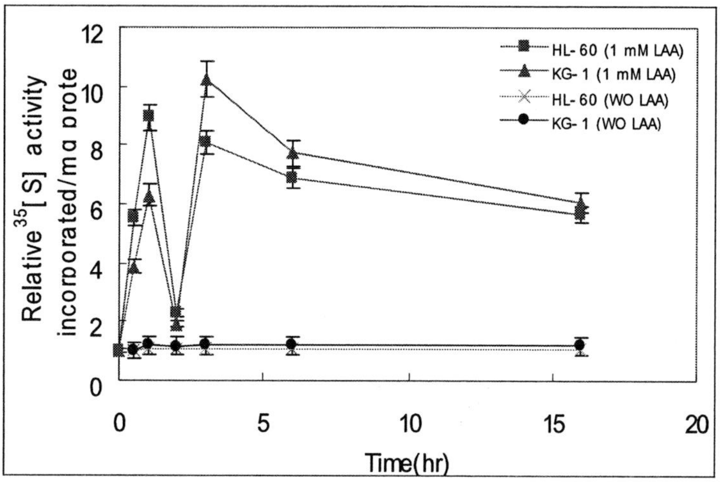

Figure 2. Effect of LAA on the ^{35}[S]-incorporation by cells during culture. Cells were cultured with 5 μCi of ^{35}[S]-cysteine for 30 mins and additionally incubated with or without LAA during the indicated culture period. Then, the cells were washed and the radioactivity of cell extract was measured using scintillation counter. Each point represents a mean value of cumulative data of four separate experiments.

Effect of LAA on GST Enzyme Activity and GST mRNA Synthesis

As expected, LAA also had stimulatory effects on the level of activity of GST in cells (235% increase in HL-60 cells and 159% increase in NB4 cells compared with the control) (Figure. 3). To examine whether the transcriptional regulation of GST genes is responsible for the increase in the GST enzyme activity of the LAA-treated HL-60 cells, RT-PCR was carried out using primers containing non-homologous regions in different genes of GST isoforms (GST-P1, A2, A4, M1, M2). From RT-PCR, the probes were prepared and effects of LAA on the expression of GST-P1, A2, A4, M1, M2 mRNA were estimated by Northern blotting. GST-P1 and A2 mRNA levels increased (unpublished data) but that GST-A4, M1, M2 mRNA levels were not affected by LAA. LAA increased the expression of GST-P1 and A2 mRNA in concentration dependent manners in HL-60 cells. The combined observation points to a stimulatory role for LAA in the expression of GST-P mRNA followed by the concomitant increase in the GST activity.

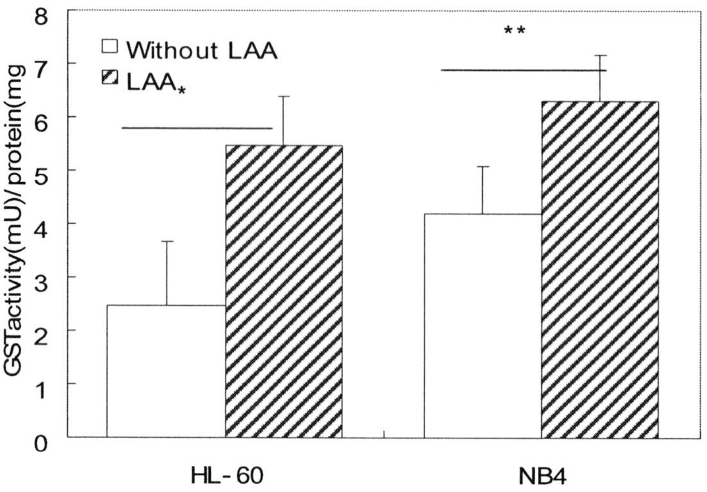

Figure 3. Effect of LAA on the activity of glutathione-S-transferase. HL-60 cells and NB4 cells were treated with 1 mM LAA for 3 hrs, and then GST activity was assayed. Briefly, 1 x 10^6 cells were harvested and washed with PBS, and freeze-thawed for lysis. GST activity was measured at 340 nm in 1-chloro-2,4-dinitrobenzene (CDNB) and GSH for 5 min. Asterisks indicate statistically significant differences (*$P<0.01$,**$P<0.05$).

Effects of LAA and Arsenic Trioxide on Intracellular ROS Accumulation

Several previous works indicated that As_2O_3 or LAA-induced apoptosis was modulated by the cellular glutathione redox system, with increased intracellular levels of reduced glutathione (GSH) having an inhibitory effect [1,15,28]. Because of the antioxidant function of GSH, it can be wondered whether modulated GSH level could trigger leukemic cell apoptosis through the generation of ROS. Jing and co-workers showed that the induction of NB4 cell apoptosis and the refractoriness of other leukemic cells to 1 μM As_2O_3 were explained, at least in part, by the accumulation of higher H_2O_2 levels in As_2O_3-treated NB4 as compared with other leukemic cells [18]. Moreover, they showed that, in the presence of glutathione peroxidase (GPx) and catalase inhibitors, it is possible to achieve high apoptotic rates in leukemic cells otherwise refractory to a therapeutic concentration of As_2O_3 (1μM). Although apoptosis was not detected in As_2O_3-treated U937 cells, cell growth was inhibited by As_2O_3 both in NB4 and U937 cells, with IC_{50} of about 0.7 and 1.2 mM at 3 days of treatment, respectively. The growth inhibition was not correlated with arrest in a specific cell cycle phase [18]. These data were consistent with the previous report that As_2O_3 inhibited cell growth in several lymphoma cells by prolongation of the cell cycle without blocking cells in specific phase [1]. Thus, the selective effect of As_2O_3 in NB4 cells was due to apoptosis induction and not to growth inhibition. They suggested that this differential sensitivity to As_2O_3-induced apoptosis in NB4 cells was associated with differences in cellular H_2O_2 levels that were determined.

A causal link between H_2O_2 levels and apoptosis is supported by the coordinate regulation of these events in response to both positive and negative regulation of H_2O_2-catabolizing enzymes.

Consistent with its proposed pro-oxidant activity, the LAA-mediated decrease in GSH/GSSG ratio correlated functionally with an intracellular increase in H_2O_2 level for the four kinds of leukemia cell lines (HL-60, NB4, NB4-R1 and KG1) ([28], Figure 1). This differential sensitivity to As_2O_3 or LAA-induced apoptosis in leukemic cells was associated with differences in cellular H_2O_2 levels that were determined by FACS analysis of cells labeled with H_2DCFDA [28,18]. After treatment with 1mM LAA, the marker area percentage representing oxidized DCF increased compared with the untreated controls from 3.7 to 19.7% in HL-60 cells, from 0.1 to 11.9% in NB4 cells, and from 1.0 to 10.4% in NB4-R1 cells within 1 h. LAA-mediated increase in H_2O_2 level was also observed in cells from patients with AML consistent with its GSH oxidation activity in these cells [28]. HL-60, NB4 or NB4-R1 cells had higher oxidized DCF contents than OVCAR-3, 2774 or SK-OV-3 cells 1 h after adding LAA. This demonstrated that the level of H_2O_2 was correlated with the apoptotic activity of LAA. Thus, both the constitutive level of H_2O_2 and the ability of As_2O_3 and LAA to increase the level of H_2O_2 were correlated with the apoptotic activity of As_2O_3 and LAA.

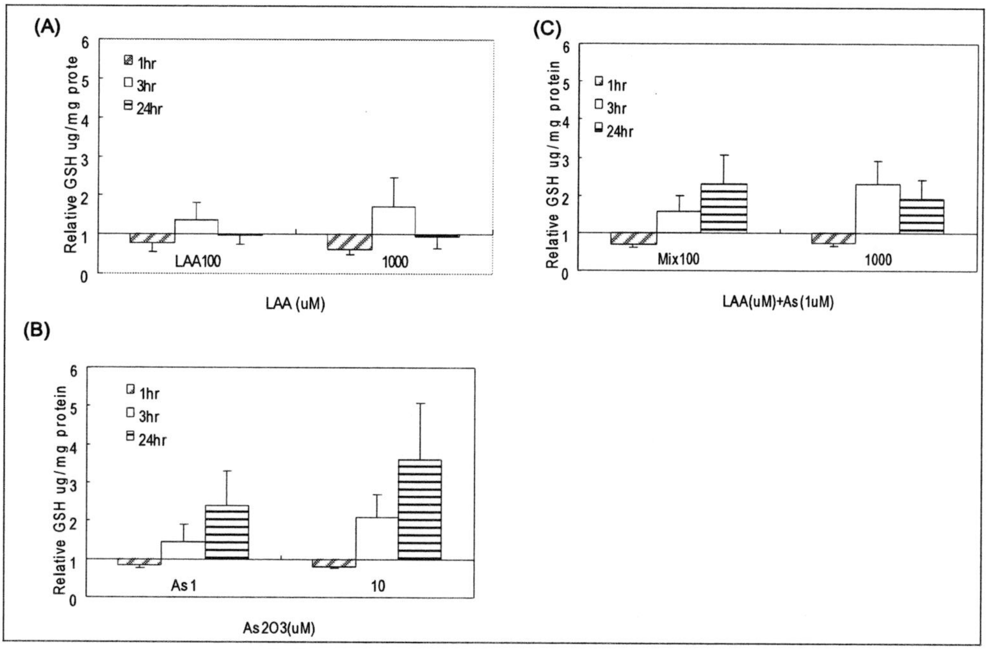

Figure 4. The combined effect of LAA and As_2O_3 on the intracellular redox system. (A) The effect of LAA on the concentration of total GSH. HL-60 cells were incubated with 0.1 and 1 mM of LAA, and changes in intracellular GSH were estimated. Values are means ± S.D. of four experiments (B) The effect of As_2O_3 on the concentration of total GSH. HL-60 cells were incubated with 1 and 10 μM of LAA, and changes in intracellular GSH were estimated. (C) The combined effect of LAA and As_2O_3 on the concentration of total GSH.

However, their characteristic role in the changes of H_2O_2 constitutive level is discriminated. As shown in Figure 1B and 4, LAA dramatically augmented intracellular GSH oxidation and ROS levels within 3 hr and in concentration dependent manners. However, the GSH oxidation and the ROS accumulation do not last for a long period of time. After 3 hr, the total amount of GSH is rather observed to be increased supposedly as a defense mechanism (Figure 1B). No more ROS accumulation can be inferred from the inclined GSH amount. The dramatic changes of intracellular oxidation state within 3 hr seem to be enough to induce apoptosis, judging from our experiment. It showed that the treatment with 1 mM LAA for only 30 min or 60 min and removal followed by could induce apoptosis in both HL-60 and NB4 cells. In the mean time, As_2O_3 resulted less ROS accumulation than LAA and it was not in concentration dependent manner. However, ROS accumulation increased up to 24 hr with long-lasting effect. It was observed that the treatment of LAA combined with As_2O_3 increases initiative ROS accumulation as well as keeping the effect high up to 24 hr (Figure 5). These results are consistent with the cellular data showing that apoptosis induced by As_2O_3 is evident in three days [1,15].

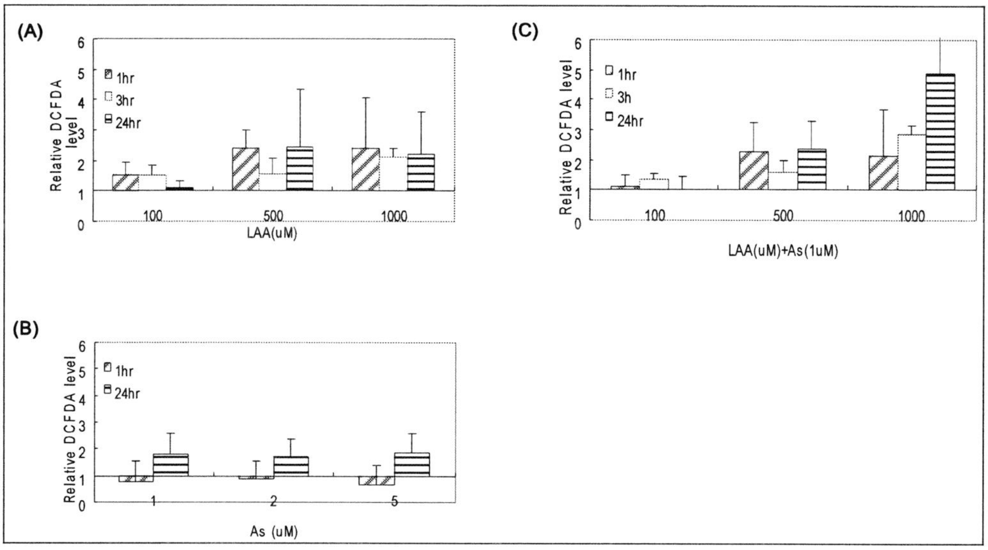

Figure 5. The combined effect of LAA and As_2O_3 on the intracellular ROS. (A) The effect of LAA on the intracellular ROS accumulation. HL-60 cells were labeled with H_2DCFDA fluorescent probe for 30 min and then treated with 0.1, 0.5 and 1 mM of LAA. Fluorescence was analyzed by flowcytometry. Values are means ± S.D. of four experiments (B) The effect of As_2O_3 on the intracellular ROS accumulation. (C) The combined effect of LAA and As_2O_3 on the intracellular ROS accumulation.

Molecular Mechanisms of Apoptosis Induced by LAA or As_2O_3

Lengthened Cell Cycle Time and Decreased Expression of PML Protein

As mentioned above, the cell cycle time was prolonged in 1 mM As_2O_3-treated Namalwa and Raji cells [1]. The cell cycle time was 35.4 hours versus 63.4 hours and 49.9 hours versus 75.3 hours for untreated and 1 mM As_2O_3-treated Namalwa and Raji cells, respectively. No substantial alteration in cell cycle time was found in Jurkat cells upon treatment with 1 mM As_2O_3 (28.1 hours), as compared with untreated cells (32.2 hours). These results suggest that 1 mM As_2O_3 can inhibit proliferation of some malignant lymphocytic cell lines by prolonging the cell cycle instead of arresting cells in a specific phase [1]. Moreover, western blotting revealed that expression of cell cycle-related proteins, including Rb, cyclin D1, CDK4, p16, p53, and c-myc, was not substantially altered in a number of cell lines after treatment with 1 mM As_2O_3 for 12, 24, and 48 hours [1]. However, the PML protein is degraded in NB4 cells treated with 1 mM As_2O_3 [12].

It was reported that As_2O_3 decreased the $\Delta\psi_m$ without altering plasma membrane permeability in SKW-3 and Namalwa cells, in which apoptosis was induced by 1 mM and 2 mM, As_2O_3 respectively [1]. It is known that $\Delta\psi_m$ collapse results in the activation of caspases, which play a central role in the apoptosis-signaling pathway [43,44]. It was found that caspase-3 precursor was cleaved into its active form in 2 mM As_2O_3-treated Namalwa and Raji cells. At the same time, its substrate PARP was also cleaved [1]. However, activation of caspase-3 and degradation of PARP were not observed in As_2O_3-treated BJAB, Molt-4, and SKW-3 cells, all of which underwent substantial apoptosis in response to 1 mM As_2O_3 [1]. Concomitant with the As_2O_3 induction of maximal cytochrome c release, Cpp32 was activated, as shown by the degradation of its precursor (32 kD) and the cleavage of PARP, a Cpp32 substrate. These events were independent of Bcl-2 degradation and were followed by DNA fragmentation [1]. From the result, it is supposed that different cell specific downstream signaling occurs during apoptosis induced by As_2O_3.

Interestingly, it was found that P388 lymphoma cells with multidrug resistance (MDR1) to adriamycin and taxol remain responsive to As_2O_3 [18,45,46]. This suggests that As_2O_3 is not a substrate for the MDR1 drug efflux pathway.

It was demonstrated that apoptosis in As_2O_3-treated NB4 cells proceeds via the classical pathway described for other ROS inductive signals [18,47-50]. Thus, As_2O_3 elicited a rapid decrease in mitochondrial membrane potential, which preceded the release of cytochrome c, Cpp32 activation, DNA fragmentation, and morphologic evidence of apoptosis [18]. However, they did not observe Bcl-2 degradation in NB4 cells treated with 1 to 2 μM As_2O_3 [18]. This is consistent with a report showing that Bcl-2 is not degraded in NB4 cells, even when treated with μM As_2O_3 [51]. Moreover, human t(14:18) B-cell lymphoma su-DHL-4 cells overexpressing Bcl-2 are sensitive to 2 μM As_2O_3 without degradation of Bcl-2 [15]. It was suggested that NIH3T342 and HL-60 cells with forced expression of Bcl-2 are more resistant to As_2O_3-induced apoptosis [18]. These findings are consistent with the result that cells overexpressing Bcl-2 are resistant to H_2O_2-induced cell death [52]. As_2O_3- induced apoptosis in NB4 cells was blocked by the caspase inhibitor Z-VAD-FMK [18] suggesting that caspase 9 and caspase 3 were involved in this pathway. This result is consistent with the previous observation that Z-VAD-FMK blocked As_2O_3-induced apoptosis in lymphoma cells [53].

Many previous reports demonstrated that free radical-induced apoptosis may involve the mitochondrial permeability transition pore and is associated with the suppression of Bcl-2 and activation of Bax [1,54]. Treatment of LAA for 24 h in HL-60 cells and for 36 h in NB4 cells repressed the constitutive expression of Bcl-2 while strongly increased the Bax expression via a dose-dependent manner [28]. This result indicated that LAA decreased the ratio of Bcl-2/Bax. It is well known that the decrease of Bcl-2/Bax ratio alters mitochondrial membrane permeability resulting in the release of cytochrome C from mitochondria into the cytosol [55]. Cytochrome C in mitochondria fraction had gradually decreased to a barely detectable level at 0.5mM and then completely disappeared at 1mM of LAA. On the contrary, cytochrome C in cytosolic fraction was increased. Caspase-9 is known to be activated by proteolytic cleavage when interacted with dATP, Apaf-1, and extra mitochondrial cytochrome C [56]. Caspase-9 can activate caspase-3 by proteolytic cleavage and the activated caspase-3 may then cleave vital cellular proteins such as PARP or activate additional caspases by proteolytic cleavage. Caspase-9 and caspase-3 were both activated through the cleavage of the inactive form of procaspase-9 and procaspase-3 via a dose-dependent manner, after being treated with LAA for 24 h in HL-60 cells and for 36 h in NB4 cells [28]. We also found that constitutive PARP was cleaved by the treatment of LAA. PARP is an enzyme involved in DNA repair and genomic integrity. 116 kDa of PARP is proteolytically cleaved into 86 kDa and 23 kDa fragments during apoptosis by caspase-3 which reduces PARP's enzymatic activity, thereby inhibiting DNA repair. These results indicated that LAA caused apoptosis by way of inducing a decrease in Bcl-2/Bax ratio, release of cytochrome C from mitochondria to cytosol, activation of caspase-9 and caspase-3, and cleavage of PARP.

It has been demonstrated that, in acute promyelocytic leukemia cells, As_2O_3 lowers bcl-2 levels and reduces NF-κB translocation to the nucleus, whereas it also induces collapse of mitochondrial transmembrane potential, resulting in cytochrome *c* release and activation of caspase-3 [57,58]. All these events appear to contribute to the induction of apoptosis, and the upstream regulatory mechanisms of such pro-apoptotic signals are recently suggested.

Recent studies showed that all-trans-retinoic acid (RA) or As_2O_3 induce activation of the p38 MAP kinase pathway, and that pharmacological inhibition of p38 augments the antiproliferative and differentiating effects of RA on NB4 cells [59,60]. These studies established that the small G-protein Rac1 and the MAPKAP kinase 2 are both RA-dependent and arsenic-dependent upstream and downstream effectors, respectively, of p38 MAP kinase. Furthermore, they indicated that treatment with a p38 inhibitor enhanced RA-dependent induction of cell proliferation and arsenic-dependent growth inhibitory responses. In our previous study, we determined that the treatment with LAA induced activation of the ERK MAP kinase pathway in HL60, NB4, and NB4-R1 acute myeloid leukemia cell lines [61]. Considering that LAA induced activation of ERK occurs both in cells that are resistant and sensitive to differentiation by RA, the upstream signals that mediate LAA-dependent activation of ERK are separated from the ones mediating RA-dependent signals. From the above results, ERK rather than p38 is activated by LAA in acute myeloid leukemia cell lines such as HL-60, NB4 and NB4-R1, indicating that it also involved in the induction of LAA-response in these cells.

It is well known that Raf1 is an upstream regulator of the ERK MAP kinase in response to stress [62]. Ras is a member of family of small GTPases and is a key mediator of intracellular signaling cascades, including the Raf pathway [63]. The Raf-dependent signaling pathways have been shown to regulate the ERKs [62]. ERK1 and ERK2 were among the first substrates of Raf to be discovered and have proven to be ubiquitous and versatile mediators of Raf signal transduction [64-67]. LAA treatment resulted in activation of Rafl in three cell lines including HL-60, NB4 and NB4-R1 [61] providing evidence for involvement of Rafl kinase in a LAA-activated signal pathway. These results suggest that Rafl is an upstream regulator of the LAA-mediated activation of ERK and that the early upstream signal induced by LAA is different from an As_2O_3-activated upstream signal utilizing Rac1 and subsequent p38 MAP kinase activation.

The pre-treatment with the chemical antioxidant, N-acetyl cysteine attenuated Raf-1 activation, suggesting that H_2O_2 may be required for the signaling pathway leading to Raf-1 activation [61]. In addition, genistein, a tyrosine kinase inhibitor, reduced the activation of Raf-1 kinase by LAA [61]. It is assumed that tyrosine phosphorylation is required for the activation of Raf-1.

Previous work has shown that MAPK-activated protein kinase 2 (MAPKAP kinase 2) is a downstream effector of p38, which mediates intracellular signals in response to growth factors, hormones and neurotransmitters [68]. In our study [61], however, LAA-induced activation of ERK promotes downstream activation of MAPKAP kinase 2, in HL-60, NB4 or NB4-R1 cells. Lysates from LAA-treated cells were immunoprecipitated with an anti-MAPKAP kinase 2 antibody, and applied to an in vitro kinase assay using peptide as exogenous substrates. LAA treatment of cells activated the MAPKAP kinase 2, evidenced by the phosphorylation of substrate peptide in the kinase assay [61]. In addition, treatment of cells with the ERK-specific inhibitor PD98059 blocked the LAA-dependent activation of MAPKAP kinase 2 [61] indicating the downstream engagement of MAPKAP kinase 2 in an LAA-activated signal transduction pathway.Although LAA slightly activated p38 by Western blot, SB203580 partially reduced the activation of MAPKAP kinase 2 [61]. Judging from our results, LAA did not induce p38 activation. Interestingly, however, MAPKAP kinase 2 was involved as a downstream modulator for both p38 and ERK signal pathway. Despite the fact that MAPKAP kinase 2 was among the specific substrates of p38 identified, its mechanism of action seems to be also regulated by ERK. One ERK-catalyzed phosphorylation site was identified in the activation loop of the C-terminal kinase of MAPKAP kinase 2 [69]. In addition, it has been observed that ERK alone induced no activation of MAPKAP kinase 2, but enhanced activation of MAPKAP kinase 2 by 3-phosphoinositide-dependent protein kinase-1 (PDK1) [68]. Thus, full activation of the MAPKAP kinase 2 may require the collaboration of protein kinases such as p38, PDK1 and ERK. Altogether, these studies suggest that treatment of cells with LAA results in sequential activation of a Rafl, ERK1/2 and MAPKAP kinase 2 pathway, whose activation is probably upstream of cell apoptosis induced by LAA. In our experiments to define the functional role of this cascade, we found that inhibition of this pathway using ERK-specific inhibitor PD98059 suppressed LAA-dependent induction of apoptosis and inhibition of cell growth [61]. Thus, it appears that important signals for regulation of LAA-dependent responses are mediated by the ERK1 and/or 2 isoforms. Consistent with this, overexpression of a dominant negative ERK mutant inhibited LAA-dependent cytotoxicity in HL-60 cells [61].

Overexpression of Raf proteins is associated with proliferation in cells, including human hematopoietic cells or human erythroid progenitor cells [70-72]. However, overexpression of activated Raf proteins is associated with cell growth arrests in some cell lines [73-75]. It is not clear why overexpression of the Raf gene can lead to such conflicting results, but it has been suggested that the opposite outcomes may be determined by the amount or activity of the Raf oncoprotein [70,76]. Under LAA exposure, strong Raf signaling results in the activation of downstream kinases including ERK and MAPKAP kinase 2, which may provide another evidence for an antiproliferative effect of not only Raf but also ERK. Our data using cells obtained freshly from three patients with AML were consistent with the previous data using cell lines, showing that LAA suppressed leukemic progenitor cell growth with this suppressive effect abrogated by an ERK inhibitor [61]. This indicated the involvement of ERK in the LAA suppressive effect in patient specimens as well as cell lines. Our current view of the apoptotic pathway induced by LAA, as supported by the abovenoted data, is depicted diagrammatically in Figure 6. Our interim conclusion is that LAA induces the apoptosis of HL-60 cells, augmented by As_2O_3, through modulation of redox status, definitely, depletion of GSH by intracellular dehydroascorbic acid reduction and consequent hydrogen peroxide accumulation. And through MAPK modulation, and induction of apoptotic cascade including Bcl2/bax, cytochrome C, caspase 9, caspase 3 and PARP.

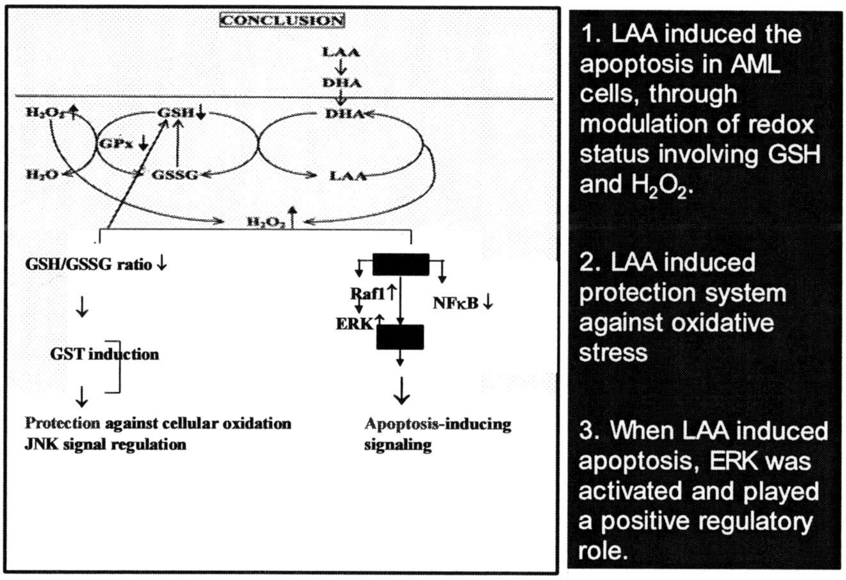

Figure 6. Diagrammatic representation of the proposed signaling pathway induced by LAA in HL-60 cells.

Inhibition of NF-κB Activation

As mentioned above, it has been demonstrated that, in acute promyelocytic leukemia cells, arsenic trioxide reduces NF-κB translocation to the nucleus [57,58]. Nuclear factor kappa-B (NF-κB) has been known to regulate expression of many genes in modulating cellular proliferation, inflammatory responses, and apoptosis [77,78]. NF-κB exists mainly as a hetero- or homo-dimer consisting of subunits of Rel family such as p50, p52, c-Rel, p65 (RelA), or RelB. NF-κB is normally sequestered in the cytoplasm as an inactive complex by binding to an inhibitory protein, IκB. IκB tightly modulates NF-κB activity by preventing nuclear translocation of NF-κB and inhibiting DNA binding. When cells are exposed to various extracellular stimuli, IkB is rapidly degraded by proteosomes following its phosphorylation and subsequent ubiquiti-nation. Hence, free NF-κB translocates to the nucleus, where it binds to its binding sites in the promoter of target genes, thereby controlling their expression [79,80].

Recently, many drugs with the ability to inhibit the cyclooxygenase (COX) enzymes have been shown to prevent or delay development of certain tumors. COX-1 is constitutively expressed to fulfill its beneficial housekeeping roles in most human tissues; whereas cyclooxygenase-2 (COX-2), with multiple binding sites on its promoter for various transcription factors including NF-κB, is not constitutively expressed in most normal tissues but it is rapidly induced by cytokines, tumor promoters, growth factor, and oncogenes [81]. The expression of COX-2 has been shown to be abnormally or overexpressed in various human tumors, including breast, head and neck, lung, pancreatic, and gastric cancers [82-86].

Unstimulated HL-60 cells had a high basal level of NF-κB activity, which was downregulated by treatment of As_2O_3, in a concentration- and time-dependent manner [29]. The NF-κB complex inactivated by As_2O_3 was indeed NF-κB existing as a heterodimer of p50 and p65 subunits. As_2O_3 inhibited NF-κB activation by preventing the degradation of IκB-α and the translocation of p65 subunit as well as inhibiting the binding between NF-κB and DNA [29]. These results are related to the data that inhibition of NF-κB strongly increases the apoptosis rate and slows down cell proliferation in Hodgkin/Reed-Sternberg cells [87]. Also, these results are important considering the data that up to 25% of hematopoietic malignancies in adults contain rearrangements or amplifications of members of the NF-κB/IκB gene families [88].

The high basal level of NF-κB activity in unstimulated HL-60 cells was almost completely downregulated by treatment of LAA, with or without 1 mM As_2O_3, in a concentration- and time-dependent manner [89]. LAA Inhibited NF-κB activation by preventing the degradation of IkB-α and the translocation of p65 Subunit but did not directly disrupt binding of NF-κB to DNA [89].

Our results indicated that inhibitory effect of As_2O_3 and LAA on NF-κB activity was dependant on GSH level confirming that intracellular GSH played a suppressive role in the inhibitory effect of As_2O_3 and LAA on NF-κB activity. Also, both As_2O_3- and LAA–induced inactivation of NF-κB activity is correlated with H_2O_2 level [29,89]. As_2O_3 and LAA inhibit expression of COX-2 and it appears to be through suppression of NF-κB DNA binding activity [29,89].

The property of As_2O_3 that disrupts the DNA binding and prevents the degradation of IkB-α and the translocation of the p65 subunit, appears to be via binding to SH groups in the DNA-binding region of NF-κB [29]. These results support the concept that intracellular GSH level could be associated with the sensitivity of cells to As_2O_3-induced apoptosis. Moreover, As_2O_3 inhibits glutatathione peroxidase activity, important in elimination of H_2O_2, and thus increases cellular H_2O_2 content in NB4 cells [90]. However, it was unclear whether LAA could bind to SH groups in the DNA-binding protein.

Early Molecular Targets Modified by L-Ascorbic Acid

To understand the molecular basis of LAA-induced oxidative stress, it is important to identify the expression or function of proteins which are modified by LAA. We analyzed molecular modifications induced by LAA in NB4 leukemia cells using a proteomic approach. This methodology provides important qualitative information on post-translational modifications to each protein and quantitative data on protein expressions in response to a particular stimulus. This information is particularly important because it provides data on early cellular events, such as, on the stimulus and signaling cascades triggered independently of protein neosynthesis.

A 30 min incubation of NB4 cells with LAA resulted in the uptake of LAA in a concentration-dependent manner (in the 0–2 mM range) [28,91]. In accordance with its proposed pro-oxidant activity, LAA-mediated reduction in the GSH/GSSG ratio correlated functionally with an intracellular H_2O_2 increase in the NB4 cell line.

The immediate effects of cell stimuli are associated with protein post-translational modifications, such as, phosphorylation, glutathionylation and cysteine oxidation. To study these early modifications, NB4 human leukemia cells were treated with 0.5 mM LAA and then collected for two-dimensional electrophoresis analysis. Approximate 240 different spots that were focused in the 4-7 pH range were analyzed per sample. After exposing cells to LAA, we observed; one new spot, three intensified spots, and five attenuated spots [91]. Each of these spots were excised, digested with trypsin and analyzed by MALDI–TOF. Peptide mass fingerprint analysis and nonredundant sequence database matching allowed the unambiguous identification of all of the analyzed species [91]. From the result, tropomyosin was identified as a marker of the oxidative effect of LAA in NB4 cells. The spot corresponding to tropomyosin was positioned at a pI value of approximately 5.0. This spot was attenuated in LAA-treated cells, in addition, a new spot was detected at pI, 4.9 of almost the same molecular weight [91]. This new spot was also identified as tropomyosin, which suggested that some chemical modification had affected its pI value. This result is consistent with the previous data that the extracellular signal-regulated kinase (ERK) mediated phosphorylation of tropomyosin-1 promotes cytoskeleton remodeling in response to oxidative stress [92]. The acidic shift of the spot with a pI 5.0 (T) to the phosphorylated tropomyosin spot by the treatment with LAA was found to be abrogated by the co-treatment with PD98059 [91], demonstrating that the phosphorylation of tropomyosin was responsible for the observed acidic shift.

The significance of this is unclear but may be related to differences in the regulation of the actomyosin contractile system in non-muscle cells as compared with that present in muscle cells. One additional point worth noting is that the proteins which specifically reacted with sera from chronic myeloid leukemia patients included structural proteins, such as, beta-tubulin and tropomyosin isoforms [31]. Although the function of these proteins in myeloid leukemia needs further investigation, tropomyosin may have value as a leukemia-associated antigen and as a molecular target in antigen-based immunotherapy. From this point of view, it is important that LAA causes a tropomyosin isoform to be modified during the immediate early response.

Another protein identified from the proteomic analysis is a thiol/disulfide exchange catalyst as a marker of the effect of LAA in NB4 cells [91]. Protein disulfide isomerase (PDI) belongs to the superfamily of thiol/disulfide exchange catalyst which acts as protein-thiol oxidoreductase enzymes and shares sequence similarities with thioredoxin [93]. PDI is composed of four domains which have similarities with thioredoxin folds (a-b-b'-a') [94]. In our study [91], the intensity of spot corresponding to the PDI b subunit was weakened in LAA-treated cells versus control cells. These results demonstrate that thiol/disulfide exchange proteins are regulated in NB4 cells after LAA exposure. This is not surprising because our study also shows that intracellular GSH/GSSG exchange occurs shortly after LAA exposure ([28], Figure 1). Therefore, we speculate that the biochemical pathway leading to thiol/disulfide redox regulation could be triggered by LAA.

Interestingly, of the proteins with expressions changed by LAA treatment, immunoglobulin-heavy-chain binding protein (BiP, identical to Hsp70 chaperone) [91] like PDI was also a multi-domain chaperone. BiP was associated with the α-subunit of prolyl 4-hydroxylase (P4-H) by a disulfide bond [90]. P4-H consists of two distinct polypeptides, the catalytically more important α-subunit and the β-subunit, which is identical to the multifunctional enzyme, PDI [90]. Thus, BiP was associated with α-subunit of P4-H, which is another partner of PDI. It was determined that the interaction of PDI with its substrates is due to a change in disulfide bonds, indicating that intra-chain disulfide bonds between domains and substrates had been reduced [91]. Taken together, these results suggest that LAA oxidizes intracellular reduced glutathione and modulates disulfide bond formation in proteins.

REFERENCES

- Zhu, X. H., Shen, Y. L., Jing, Y. K., Cai, X., Jia, P. M., Huang, Y., Tang, W., Shi, G. Y., Sun, Y. P., Dai, J., Wang, Z. Y., Chen, S. J., Zhang, T. D., Waxman, S., Chen, Z., Chen, G. Q. (1999). Apoptosis and Growth Inhibition in Malignant Lymphocytes After Treatment With Arsenic Trioxide at Clinically Achievable Concentrations. *J Natl Cancer Inst, 91*, 772-778.
- Rew, D. A. (1988). Cell and molecular mechanisms of pathogenesis and treatment of cancer. *Postgrad Med J, 74*, 77-88.
- Steller, H. (1995). Mechanisms and geges of cellular suicide. *Science, 267*, 1445-1449.
- Campana, D., Coustan-Smith, E., Manabe, A., Buschle, M., Raimondi, S. C., Behm, F. G., et al. (1993). Prolonged survival of B-lineage acute lymphoblastic leukemia cells is accompanied by overexpression of bcl-2 protein. *Blood, 81*, 1025-1031.
- Coustan-smith, E., Kitanaka, A., Pui, C. H., Nc-Ninch, L., Evans, W. E., Raimindi, S. C., et al. (1996). Clinical relevance of BCL-2 overexpression in childhood acute lymphoblastic leukemia. *Blood, 87*, 1140-1146.
- Kitada, S., Andersen, J., Akar, S., Zapata, J. M., Takayama, S., Krajewski, S., et al. (1998). Expression of apoptosis-regulating proteins in chronic lymphocytic

leukemia: correlations with *in vitro* and *in vivo* chemoresponses. *Blood, 91*, 3379-3389.

- Fisher, D. E. (1994). Apoptosis in cancer therapy: crossing the threshold. *Cell, 78*, 539-542.
- Staunton, M. J., Gaffney, E. F. (1998). Apoptosis: basic concepts and potential significance in human cancer. *Arch Pathol Lab Med, 122*, 310-319.
- Sun, H. D., Ma, L., Hu, X. C., Zhang, T. D. (1992). Ai-Lin I treated 32 cases of acute promyelocytic leukemia. *Chin J Integrat Chinese Western Med, 12*, 170-172.
- Zhang, P., Wang, S. Y., Hu, X. H. (1996). Arsenic trioxide treated 72 cases of acute promyelocytic leukemia. *Chin J Hematol. 17*, 58-61.
- Mervis, J. (1996). Ancient remedy performs new tricks[news]. *Science, 273*, 578.
- Chen, G. Q., Zhu, J., Shi, X. G., Ni, J. H., Zhong, H.J., Si, G.Y., Jin, X. L., Tang, W., Li, X. S., Xong, S. M., Shen, Z. X., Sun, G. L., Ma, J., Zhang, P., Zhang, T. D., Gazin, C., Naoe, T., Chen, S. J., Wang, Z. Y., Chen, Z. (1996). *In vitro* studies on cellular and molecular mechanisms of arsenic trioxide (As_2O_3) in the treatment of acute promyelocytic leukemia: As_2O_3 induces NB4 cell apoptosis with downregulation of Bcl-2 expression and modulation of PML-RAR alpha/PML proteins. *Blood, 88*, 1052-1061.
- Chen, G. Q., Shi, X. G., Tang, W., Xiong, S. M., Zhu, J., Cai, X., et al. (1997). Use of arsenic trioxide (As_2O_3) in the treatment of acute promyelocytic leukemia (APL): I. As_2O_3 exerts dose-dependent dual effects on APL cells. *Blood, 89*, 3345-3353.
- Kitamura, K., Yoshida, H., Ohno, R., Naoe, T. (1997). Toxic effects of arsenic (As^{3+}) and other metal ions on acute promyelocytic leukemia cells. *Int J Hematol, 65*, 179-185.
- Dai, J., Weinberg, R. S., Waxman, S., Jing, Y. (1999). Malignant cells can be sensitized to undergo growth inhibition and apoptosis by arsenic trioxide through modulation of the glutathione redox system. *Blood, 93*, 268-277.
- Bahlis, N: J., Jordan-Mcmurry, I., Grad, J. M., Reis, I., Neel, J., Kharfan-Dabaja, M. et al. (2001). Phase I results from a phase I/II study of arsenic trioxide (As_2O_3) and ascorbic acid (AA) in relapsed and chemorefractory multiple myeloma. *Blood, 98*, 375a.
- Grad, J. M., Bahlis, N. J., Reis, I., Oshiro, M. M., Dalton, W. S., Boise, L. H. (2001). Ascorbic acid enhances arsenic trioxide-induced cytotoxicity in multiple myeloma cells. *Blood, 98*, 805-813.
- Jing, Y., Dai, J., Chalmers-Redman, R. M. E., Tatton, W. G., Waxman, S.(1999). Arsenic trioxide selectively induces acute promyelocytic leukemia cell apoptosis via a hydrogen peroxide-dependent pathway. *Blood, 94*, 2102-2111.
- Bram, S., Froussard, P., Guichard, M. (1980). Ascorbic acid preferential toxicity for malignant melanoma cells. *Nature, 284*, 629-631.
- Bruchelt, G., Baader, L., Reith, A. G., Holger, N. L., Gebhardt, S., Niethammer, D. (1993). Rationale for the use of ascorbic acid in neuroblastoma therapy. In M. Schwab (Ed.), *Human neuroblastoma* (pp. 34-40). Harwood Academic Publishers.
- Fujinaga, S., Sakagami, H., Kuribayashi, N., Takahashi, H., Amano, Y., Sakagami, T. et al. (1994). Possible role of hydrogen peroxide in apoptosis induction by

ascorbic acid in human myelogenous leukemic cell lines. *Showa University Medical Science, 6*, 135-144.

- De Laurenzi, V., Melino, G., Savini, I., Annicchiarico-Petruzzelli, M., Finazzi-Agro, A., Avigliano, L. (1995). Cell death by oxidative stress and ascorbic acid regeneration in human neuroectodermal cell lines. *Eur J Cancer, 31*, 463-466.
- Park, C. H., Bergsagel, D. E., McCulloch, E. A. (1971). Ascorbic acid: a culture requirement for colony formation by mouse plasmacytoma cells. *Science, 174*, 720-722.
- Park, C. H. (1985). The biological nature of the effect of ascorbic acids on the growth of human leukemic cells. *Cancer Res, 45*, 3969-3973.
- Park, C. H., Kimler, B. F., Bodensteiner, D., Lynch, S. R., Hassanein, R. S. (1992). In vitro growth modulation by L-ascorbic acid of colony-forming cells from bone marrow of patients with myelodysplastic syndromes. *Cancer Res, 52*, 4458-4466.
- Sakagami, H., Asano, K., Fukuchi, K., Gomi, K., Ota, H., Kazama, K. et al. (1991). Induction of tumor degeneration by sodium benzylideneascorbate. *Anticancer Res, 11*, 1533-1538.
- Park, C. H., Kim, W. S., Park, C., Lee, M. H., Boo, Y. C., Yoon, S. S. (1999). Clinical disease suppression and reduction in acute myeloid leukemia and solid tumors by very high dose of L-ascorbic acid: A new concept and in search of molecular targets. *Clin Cancer Res, 5 (SUPPL.)*, 3784s
- Park, S., Han, S. S., Park, C. H., Hahm, E. R., Lee, S. J., Park, H. K., Lee, S. H., Kim, W. S., Jung, C. W., Park, K., Riordan, H. D., Kimler, B. F., Kim, K., Lee, J. H. (2004). L-Ascorbic acid induces apoptosis in acute myeloid leukemia cells via hydrogen peroxide-mediated mechanisms. *Int J Biochem Cell Biol, 36*, 2180-2195.
- Han, S. S., Kim, K., Hahm, E. R., Park, C. H., Kimler, B. F., Lee, S. J., Lee, S. H., Kim, W. S., Jung, C. W., Park, K., Kim, J., Yoon, S. S., Lee, J. H., Park, S. (2005). Arsenic trioxide represses constitutive activation of NF-kappaB and COX-2 expression in human acute myeloid leukemia, HL-60. *J Cell Biochem, 94*, 695-707.
- Meister, A. (1994). Glutathione-ascorbic acid antioxidant system in animals. *J Biol Chem. 269*, 9397-9400.
- Zou, L., Wu, Y., Pei, L., Zhong, D., Gen, M., Zhao, T., Wu, J., Ni, B., Mou, Z., Han, J., Chen, Y., Zhi, Y. (2005). Identification of leukemia-associated antigens in chronic myeloid leukemia by proteomic analysis. *Leuk Res, 29*, 1387-1391.
- Lee, T.C., Wei, M. L., Chang, W. J., Ho, I. C., Lo, J. F., Jan. K. Y., Huang, H. (1989). Elevation of glutathione levels and glutathione S-transferase activity in arsenic-resistant Chinese hamster ovary cells. *In Vitro Cell Dev Biol, 25*, 442.
- Lee, T. C., Ko, J. L., Jan, K. Y. (1989). Differential cytotoxicity of sodium arsenite in human fibroblasts and Chinese hamster ovary cells. *Toxicology, 56*, 289.
- Ochi, T., Kaise, T., Oya-Ohta, Y. (1994). Glutathione plays different roles in the induction of the cytotoxic effects of inorganic and organic arsenic compounds in cultured BALB/c 3T3 cells. *Experientia, 50*, 115.
- Ochi, T., Nakajima, F., Sakurai, T., Kaise, T., Oya-Ohta, Y. (1996). Dimethylarsinic acid causes apoptosis in HL-60 cells via interaction with glutathione. *Arch Toxicol, 70*, 815.

- Scott, N., Hatlelid, K. M., MacKenzie, N. E., Carter, D. E. (1993). Reactions of arsenic(III) and arsenic(V) species with glutathione. *Chem Res Toxicol, 6*, 102.
- Wang, T. S., Kuo, C. F., Jan, K. Y., Huang, H. (1996). Arsenite induces apoptosis in chinese hamster ovary cells by generation of reactive oxygen species. *J Cell Physiol, 169*, 256.
- Cantoni, O., Hussain, S., Guidarelli, A., Cattabeni, F. (1994). Cross-resistance to heavy metals in hydrogen peroxide-resistant CHO cell varients. *Mutat Res, 324*, 1.
- Wyllie, S., Fairlamb, A. H. (2006). Differential toxicity of antimonial compounds and their effects on glutathione homeostasis in a human leukaemia monocyte cell line. *Biochem Pharmacol, 71(3)*,257-267.
- Droge, W., Eck, H. P., Mihm, S. (1992). HIV-induced cysteine deficiency and T-cell dysfunctionA rationale for treatment with N-acetylcysteine. *Immunol Today, 13*, 211.
- Han, D., Handelman, G., Marcocci, L., Sen, C. K., Roy, S., Kobuchi, H., Tritschler, H. J., Flohe, L., Packer, L. (1997). Lipoic acid increases de novo synthesis of cellular glutathione by improving cystine utilization. *Biofactors, 6*, 321.
- Watanabe, H., Bannai, S. (1987). Induction of cystine transport activity in mouse peritoneal macrophages. *J Exp Med, 165*, 628-640.
- Kumar, S. (1995). ICE-like protease in apoptosis. *Trends Biochem Sci, 20*, 198-204.
- Takahashi, A., Earnshaw, W. C. (1996). ICE-related proteases in apoptosis. *Curr Opin Genet Dev, 6*, 50-55.
- Wang, H. F., Lee, T. C. (1993). Glutathione S-transferase pi facilitates the excretion of arsenic from arsenic-resistant Chinese hamster ovary cells. *Biochem Biophys Res Commun, 192*, 1093.
- Rappa, G., Lorico, A., Flavell, R. A., Sartorelli, A. C. (1997). Evidence that the multidrug resistance protein (MRP) functions as a co-transporter of glutathione and natural product toxins. *Cancer Res, 57*, 5232.
- Marchetti, P., Decaudin, D., Macho, A., Zamzami, N., Hirsch, T., Susin, S. A., Kroemer, G. (1997). Redox regulation of apoptosis: Impact of thiol oxidation status on mitochondrial function. *Eur J Immunol, 27*, 289.
- Satoh, T., Enokido, Y., Aoshima, H., Uchiyama, Y., Hatanaka, H. (1997). Changes in mitochondrial membrane potential during oxidative stress-induced apoptosis in PC12 cells. *J Neurosci Res, 50*, 413.
- Yang, J. C., Cortopassi, G. A. (1998). Induction of the mitochondrial permeability transition causes release of the apoptogenic factor cytochrome c. *Free Radic Biol Med, 24*, 624.
- Stridh, H., Kimland, M., Jones, D. P., Orrenius, S., Hampton, M. B. (1998). Cytochrome c release and caspase activation in hydrogen peroxide- and tributyltin-induced apoptosis. *FEBS Lett, 429*, 351.
- Ma, D. C., Sun, Y. H., Chang, K. Z., Ma, X. F., Huang, S. L., Bai, Y. H., Kang, J., Liu, Y. G., Chu, J. J. (1998). Selective induction of apoptosis of NB4 cells from G2+M phase bysodium arsenite at lower doses. *Eur J Haematol, 61*,27.

- Hockenbery, D. M., Oltvai, Z. N., Yin, X. M., Milliman, C. L., Korsmeyer, S. J. (1993). Bcl-2 functions in an antioxidant pathway to prevent apoptosis. *Cell, 75*, 241.
- Akao, Y., Mizoguchi, H., Kojima, S., Naoe, T., Ohishi, N., Yagi, K. (1998). Arsenic induces apoptosis in B-cell leukaemic cell lines in vitro: Activation of caspases and down-regulation of Bcl-2 protein. *Br J Haematol, 102*, 1055.
- Kroemer, G., Zamzami, N., Susin, S. A. (1997). Mitochondrial control of apoptosis. *Immunol Today, 18*, 44-51.
- Kluck, R. M., Martin, S. J., Hoffman, B. M., Zhou, J. S., Green, D. R., Newmeyer, D. D. (1997). Cytochrome c activation of CPP32-like proteolysis plays a critical role in a Xenopus cell-free apoptosis system. *EMBO Journal, 16 (15)*, 4639-4649.
- Li, P., Nijhawan, D., Budihardjo, I., Srinivasula, S. M., Ahmad, M., Alnemri, E. S., Wang, X. (1997). Cytochrome c and dATP-dependent formation of Apaf-1/caspase-9 complex initiates an apoptotic protease cascade. *Cell, 91*, 479-489.
- Wang, Z. G., Rivi, R., Delva, L., Konig, A., Scheinberg, D. A., Gambacorti-Passerini, C., Gabrilove, J. L., Warrell, R. P., Jr., and Pandolfi, P. P. (1998). Arsenic Trioxide and Melarsoprol Induce Programmed Cell Death in Myeloid Leukemia Cell Lines and Function in a PML and PML-RAR Independent Manner. *Blood, 92*, 1497-1504.
- Mahieux, R., Pise-Masison, C., Gessain, A., Brady, J. N., Olivier, R., Perret, E., Misteli, T., and Nicot, C. (2001). Arsenic trioxide induces apoptosis in human T-cell leukemia virus type 1- and type 2-infected cells by a caspase-3-dependent mechanism involving Bcl-2 cleavage. *Blood, 98*, 3762-3769.
- Alsayed, Y., Uddin, S., Mahmud, N., Lekmine, F., Kalvakolanu, D. V., Minucci, S., Bokoch, G., Platanias, L. C. (2001). Activation of Rac1 and the p38 Mitogen-activated Protein Kinase Pathway in Response to All-trans-retinoic Acid. *J Biol Chem, 276*, 4012-4019.
- Verma, A., Mohindru, M., Deb, D. K., Sassano, A., Kambhampati, S., Ravandi, F., Minucci, S., Kalvakolanu, D. V., Platanias, L. C. (2002). Activation of Rac1 and the p38 mitogen-activated protein kinase pathway in response to arsenic trioxide. *J Biol Chem, 277*, 44988-44995.
- Park, S., Park, C. H., Hahm, E. R., Kim, K., Kimler, B. F., Lee, S. J., Park, H. K., Lee, S. H., Kim, W. S., Jung, C. W., Park, K., Riordan, H. D., Lee, J. H. (2005). Activation of Raf1 and the ERK pathway in response to l-ascorbic acid in acute myeloid leukemia cells. *Cell Signal, 17*, 111-119.
- English, J., Pearson, G., Wilsbacher, J., Swantek, J., Karandikar, M., Xu, S., Cobb, M. H. (1999). New insights into the control of MAP kinase pathways. *Exp Cell Res, 253*, 255-270.
- Malumbres, M., Pellicer, A. (1998). RAS pathways to cell cycle control and cell transformation. *Front Biosci, 3*, 887-912.
- Jones, S. W., Erikson, E., Blenis, J., Maller, J. L., Erikson, R. L. (1988). A Xenopus ribosomal protein S6 kinase has two apparent kinase domains that are each similar to distinct protein kinases. *Proc Natl Acad Sci USA, 85*, 3377-3381.

- Sturgill, T. W., Ray, L. B., Erikson, E., Maller, J. L. (1988). Insulin-stimulated MAP-2 kinase phosphorylates and activates ribosomal protein S6 kinase II. *Nature, 334*, 715-718.
- Alcorta, D. A., Crews, C. M., Sweet, L. J., Bankston, L., Jones, S. W., Erikson, R. L. (1989). Sequence and expression of chicken and mouse rsk: Homologs of Xenopus laevis ribosomal S6 kinase. *Mol Cell Biol, 9*, 3850-3859.
- Zhao, A., Li, J., Lu, Y. (1995). Expression of Ras and erbB in benign and malignant lesions of the stomach. *Chinese Journal of Oncology, 17*, 336-339.
- Frodin, M., Gammeltoft, S. (1999). Role and regulation of 90 kDa ribosomal S6 kinase (RSK) in signal transduction. *Mol Cell Endocrinol, 151*, 65-77.
- Sutherland, C., Campbell, D. G., Cohen, P. (1993). Identification of insulin-stimulated protein kinase-1 as the rabbit equivalent of rsk(mo)-2. Identification of two threonines phosphorylated during activation by mitogen-activated protein kinase. *Eur J Biochem, 212*, 581-588.
- Lee J. T., McCubery, J. A. (2002). The Raf/MEK/ERK signal transduction cascade as a target for chemotherapeutic intervention in leukemia. *Leukemia,16*, 486-507.
- Hochhaus, A., McCubrey, J., Killmann, N. M. B. (2002). Molecular targets for therapy, signal transduction inhibition and imatinib. *Leukemia, 16*, 1205-1206.
- McGlynn, A. P., Padua, R. A., Burnett, A. K., Darley, R. L. (2000). Alternative effects of RAS and RAF oncogenes on the proliferation and apoptosis of factor-dependent FDC-P1 cells. *Leuk Res, 24*, 47-54.
- Cioffi, C. L., Garay, M., Johnston, J. F., McGraw, K., Boggs, R. T., Hreniuk, D., Monia, B. P. (1997). Selective inhibition of A-Raf and C-Raf mRNA expression by antisense oligodeoxynucleotides in rat vascular smooth muscle cells: Role of A-Raf and C-Raf in serum-induced proliferation. *Mol Pharmacol, 51*, 383-389.
- Ravi, R. K., Weber, E., McMahon, M., Williams, J. R., Baylin, S., Mal, A., Harter, M. L., Dillehay, L. E., Claudio, P. P., Giordano, A., Nelkin, B. D., Mabry, M. (1998). Activated Raf-1 causes growth arrest in human small cell lung cancer cells. *J Clin Invest, 101*, 153-159.
- Lloyd, A. C., Obermüller, F., Staddon, S., Barth, C. F., McMahon, M., Land, H. (1997). Cooperating oncogenes converge to regulate cyclin/cdk complexes. *Genes Dev, 11*, 663-677.
- Woods, D., Parry, D., Cherwinski, H., Bosch, E., Lees, E., McMahon, M. (1997) Raf-induced proliferation or cell cycle arrest is determined by the level of Raf activity with arrest mediated by p21(Cip1). *Mol Cell Biol, 17*, 5598-5611.
- Sen, P. K., Packer, L. (1996). Antioxidant and redox regulation of gene transcription. *FASEB J, 10*,709-720.
- Barnes, P. J., Karin, M. (1997). Nuclear factor-kappaB: a pivotal transcription factor in chronic inflammatory diseases. *N Engl J Med, 336*, 1066-1071.
- Beg, A. A., Baltimore, D. (1996). An essential role for NF-kappaB in preventing TNF-alpha-induced cell death. *Science, 274*, 782-784.
- Grimm, S., Bauer, M. K., Baeuerle, P. A., Schulze-Osthoff, K. (1996). Bcl-2 down-regulates the activity of transcription factor NF-kappaB induced upon apoptosis. *J Cell Biol, 134*, 13-23.

- Prescott, S. M., Fitzpatrick, F. A. (2000). Cyclooxygenase-2 and carcinogenesis. *Biochim Biophys Acta, 1470*, M69-78.
- Subbaramaiah, K., Telang, N., Ramonetti, J. T., Araki, R., DeVito, B., Weksler, B. B., Dannenberg, A. J. (1996). Transcription of cyclooxygenase-2 is enhanced in transformed mammary epithelial cells. *Cancer Res, 56*, 4424-4429.
- Ristimaki, A., Honkanen, N., Jankala, H., Sipponen, P., Harkonen, M. (1997). Expression of cyclooxygenase-2 in human gastric carcinoma. *Cancer Res, 57*, 1276-1280.
- Shen, Z. X., Chen, G. Q., Ni, J. H., Li, X. S., Xiong, S. M., Qiu, Q. Y., Zhu, J., Tang, W., Sun, G. L., Yang, K. Q., Chen, Y., Zhou, L., Fang, Z. W., Wang, Y. T., Ma, J., Zhang, P., Zhang, T. D., Chen, S. J., Chen, Z., Wang, Z. Y. (1997). Use of arsenic trioxide (As_2O_3) in the treatment of acute promyelocytic leukemia (APL): II. Clinical efficacy and pharmacokinetics in relapsed patients. *Blood, 89*, 3354-3360.
- Chan, G., Boyle, J. O., Yang, E. K., Zhang, F., Sacks, P. G., Shah, J. P., Edelstein, D., Soslow, R. A., Koki, A. T., Woerner, B. M., Masferrer, J. L., Dannenberg, A. J. (1999). Cyclooxygenase-2 expression is up-regulated in squamous cell carcinoma of the head and neck. *Cancer Res, 59*, 991-994.
- Tucker, O. N., Dannenberg, A. J., Yang, E. K., Zhang, F., Teng, L., Daly, J. M., Soslow, R. A., Masferrer, J. L., Woerner, B. M., Koki, A. T., Fahey, T. J. 3rd. (1999). Cyclooxygenase-2 expression is up-regulated in human pancreatic cancer. *Cancer Res, 59*, 987-990.
- Bargou, R. C., Emmerich, F., Krappmann, D., Bommert, K., Mapara, M. Y., Arnold, W., Royer, H. D., Grinstein, E., Greiner, A., Scheidereit, C., Dorken, B. (1997). Constitutive nuclear factor-kappaB-RelA activation is required for proliferation and survival of Hodgkin's disease tumor cells. *J Clin Invest, 100*, 2961-2969.
- Luque, I., Gelinas, C. (1997). Rel/NF-kappa B and I kappa B factors in oncogenesis. *Semin Cancer Biol, 8*, 103-111.
- Han, S. S., Kim, K., Hahm, E. R., Lee, S. J., Surh, Y. J., Park, H. K., Kim, W. S., Jung, C. W., Lee, M. H., Park, K., Yang , J. H., Yoon, S. S., Riordan, N. H., Riordan, H. D., Kimler, B. F., Park, C. H., Lee, J. H., Park, S. (2004). L-ascorbic acid represses constitutive activation of NF-kappaB and COX-2 expression in human acute myeloid leukemia, HL-60. *J Cell Biochem, 93*, 257-270.
- John, D. C., Bulleid, N. J. (1996). Intracellular dissociation and reassembly of prolyl 4-hydroxylase: the alpha-subunits associated with the immunoglobulin-heavy-chain binding protein (BiP) allowing reassembly with the beta-subunit. *Biochem J, 317*, 659-665.
- Park, S., Lee, J., Yeom, C. H. (2006). A proteomic approach to the identification of early molecular targets changed by L-ascorbic acid in NB4 human leukemia cells. *J Cell Biochem, published online Jul 5* 2006, DOI:10,1002/jcb20971.
- Houle, F., Rousseau, S., Morrice, N., Luc, M., Mongrain, S., Turner, C. E., Tanaka, S., Moreau, P., Huot, J. (2003). Extracellular signal-regulated kinase mediates phosphorylation of tropomyosin-1 to promote cytoskeleton remodeling in response to oxidative stress: impact on membrane blebbing. *Mol Biol Cell, 14*, 1418-1432.

- Noiva, R. (1999). Protein disulfide isomerase: the multifunctional redox chaperone of the endoplasmic reticulum. *Semin Cell Dev Biol, 10*, 481-493.
- Kemmink, J., Darby, N. J., Dijkstra, K., Nilges, M., Creighton, T. E. (1997). The folding catalyst protein disulfide isomerase is constructed of active and inactive thioredoxin modules. *Curr Biol, 7*, 239-245.

In: Cell Apoptosis and Cancer
Editor: Albina W. Taylor, pp. 113-133
ISBN: 978-1-60021-506-3

Chapter VII

APOPTOSIS-INDUCING EFFECTS OF CURCUMIN IN ADULT T-CELL LEUKEMIA CELLS

Mariko Tomita and Naoki Mori*
Division of Molecular Virology and Oncology, Graduate School of Medicine, University of the Ryukyus, Nishihara, Japan.

ABSTRACT

Adult T-cell leukemia (ATL) is a fatal malignancy of T lymphocytes caused by infection with human T-cell leukemia virus type I (HTLV-I) and remains incurable. Leukemic cells often evade apoptosis by expressing several anti-apoptotic proteins, down-regulation and mutation of proapoptotic genes, and alterations in signaling pathways that give them survival advantage and thereby allow them to resist therapy-induced apoptosis. Multiple oncogenic signals such as NF-κB, AP-1, phosphatidylinositol 3-kinase (PI3K)-Akt, and Janus kinase (JAK)-signal transducer and activator of transcription (STAT) signaling pathways are constitutively activated in HTLV-I-infected T cells, suggesting possible involvement of these signaling pathways in anti-apoptosis and leukemogenesis of ATL. Curcumin (diferuloylmethane), a natural yellow compound present in turmeric (*Curcuma longa* Linn), has been studied vigorously as a potent chemopreventive agent for cancer because of its inhibitory effect on proliferation and induction of apoptosis in several tumor cell lines. Curcumin prevents the growth of HTLV-I-infected T-cell lines and primary ATL cells by inducing apoptosis. Curcumin also inhibits the growth of HTLV-I-infected T-cell tumors implanted subcutaneously in SCID mice. These effects of curcumin are mediated by inhibiting the NF-κB, AP-1, PI3K-Akt, and JAK-STAT signaling pathways. The ability of curcumin to alter gene transcription and induce apoptosis in preclinical models is likely to be of particular relevance to chemoprevention and chemotherapy in ATL

* Correspondence concerning this article should be addressed to Prof. Naoki Mori, MD, PhD, Division of Molecular Virology and Oncology, Graduate School of Medicine, University of the Ryukyus, 207 Uehara, Nishihara, Okinawa 903-0215, Japan. Phone: +81-98-895-1130. Fax: +81-98-895-1410. E-mail: n-mori@med.u-ryukyu.ac.jp.

patients. This review describes the molecular mechanisms of curcumin-induced apoptosis in HTLV-I-infected T-cell lines and primary ATL cells.

Keywords: HTLV-I, ATL, curcumin, apoptosis, NF-κB, AP-1, PI3K, Akt, JAK, STAT

Abbreviations

AP-1, activator protein 1;
ATL, adult T-cell leukemia;
HTLV-I, human T-cell leukemia virus type I;
IAP, inhibitor of apoptosis protein;
IκB, inhibitor of NF-κB;
IKK, IκB kinase;
IL-2Rα, interleukin-2 receptor α chain;
JAK, Janus kinase;
PBMCs, peripheral blood mononuclear cells;
PDK, phosphoinositide-dependent kinase;
RHD, Rel homology domain;
PI3K, phosphatidylinositol 3-kinase;
PIP_2, phosphatidylinositol-4,5-bisphosphate;
PIP_3, phosphatidylinositlo-3,4,5-triphosphate;
SHP, Src homology phosphatase;
SOCS, suppressors of cytokine signaling;
STAT, signal transducer and activator of transcription.

Introduction

Human T-cell leukemia virus type I (HTLV-I) is the etiologic agent for adult T-cell leukemia (ATL), a severe and fatal malignancy of mature $CD4^+$ T cells. ATL generally occurs in individuals infected around the time of birth. While the vast majority of infected individuals remain clinically asymptomatic, only 2-5% will develop ATL after a prolonged latent period ranging from 20 to 60 years, consistent with an age-related accumulation of leukemogenic events and a multistep process of transformation. ATL is subclassified into four subtypes: acute, lymphoma, chronic, and smoldering types. In the relatively indolent chronic and smoldering types, the median survival time is ≥2 years. However, regardless of intensive chemotherapies, the prognosis of acute and lymphoma types ATL has not improved. The condition often progresses to death with a median survival time of 13 months in these types ATL with the most intensive chemotherapy [1]. Such poor prognosis might be due to the resistance of ATL cells to apoptosis induced by anti-cancer drugs. Leukemic cells often evade apoptosis by expressing several anti-apoptotic proteins, down-regulation and mutation of proapoptotic genes, and alterations in signaling pathways that give them survival advantage and thereby allow them to resist therapy-induced apoptosis.

The HTLV-I carries a single-stranded RNA genome which encodes the structural and enzymatic proteins, Gag, Env, and Pol, similar to other retroviruses. In addition to the viral structural genes, the HTLV-I genome contains a unique region at the 3' end, designated the pX region open reading frames, which encodes a number of nonstructural proteins, including Tax, Rex, and the accessory proteins. Although the exact mechanism of leukemogenesis is unknown at present, Tax, a 40-kDa transactivator protein comprised of 353 amino acid residues, is considered to play a crucial role in disease development [2]. In addition to regulating viral gene expression through the 5' long terminal repeat, Tax induces or represses the expression of a large variety of cellular genes. Lacking a typical DNA-binding domain, Tax does not have the capacity to transactivate target genes directly. However, it can modulate the activity of a number of transcription factors by activating different signaling pathways, such as NF-κB, activator protein 1 (AP-1), phosphatidylinositol 3-kinase (PI3K)-Akt, and Janus kinase (JAK)-signal transducer and activator of transcription (STAT), which have been demonstrated to be involved in leukemogenesis of ATL [2].

Curcumin (diferuloylmethane), the major yellow pigment in turmeric, which is commonly used as a flavoring and coloring agent in foods, drinks, and cosmetics, is obtained from rhizomes of the plant *Curcuma longa* Linn. Several recent studies have shown that curcumin has anti-inflammatory, anti-oxidant, and anti-carcinogenic properties [3]. Interest in the anti-cancer actions of curcumin is considerable, and research efforts are expanding in this area. Curcumin inhibits cell proliferation and induces apoptosis in a variety of human cancer cell lines, including breast, prostate, colon, oral squamous carcinoma, melanoma, lymphoma, and leukemia cell lines [3]. It has been reported that curcumin induces apoptosis through mitochondrial pathway involving caspase-8, BID cleavage, cytochrome *c* release, and caspase-3 activation in a myeloid leukemia cell lines [4]. Bcl-2 and Bcl-x_L are also critical negative regulators of curcumin-induced apoptosis; curcumin suppresses the constitutive expression of Bcl-2 and Bcl-x_L in mantle cell lymphoma [5] and multiple myeloma [6] cell lines. Curcumin also activates caspase-7 and caspase-9 and induces PARP cleavage in both cell lines [5,6]. The effects of curcumin on induction of apoptosis are by no means consistent, and they may be tissue-specific. Pharmacologically, curcumin is safe; preclinical studies of curcumin showed no toxicity from 1.2 g/kg body weight curcumin administered to rats for 14 days [7] or from 300 mg/kg body weight curcumin administered to mice for 14 weeks [8]. Human clinical trials indicated no dose-limiting toxicity when administered at doses up to 3.6 g/day [9]. All of these studies suggest that curcumin has enormous potential in the prevention and therapy of cancer.

This review will summarize the effects of curcumin on growth and survival of HTLV-I-infected T-cell lines and primary ATL cells, and describe the molecular mechanisms of curcumin-induced apoptotic effects on these cells by focusing on the signaling pathways, such as NF-κB, AP-1, PI3K-Akt, and JAK-STAT, which are involved in leukemogenesis of ATL.

Curcumin Inhibits Growth of HTLV-I-Infected T-Cell Lines and Primary ATL Cells by Inducing Apoptosis

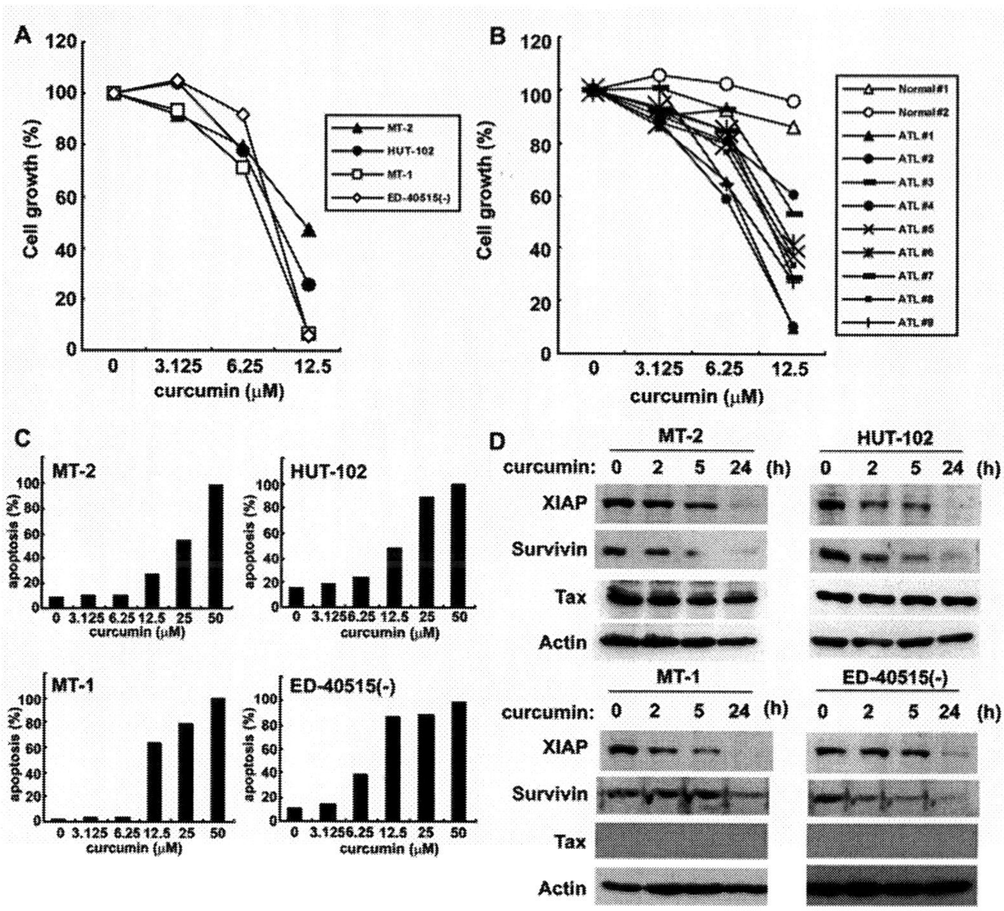

Figure 1. Curcumin induces apoptosis of HTLV-I-infected T cells. HTLV-I-infected T-cell lines [MT-2, HUT-102, MT-1, and ED-40515(-)] (A) and PBMCs from ATL patients (ATL) and healthy donors (Normal) (B) were treated with the indicated concentrations of curcumin for 48 h. Cell growth was assessed by the WST-8 method. (C) Induction of apoptosis in HTLV-I-infected T-cell lines by curcumin. Cells were treated with the indicated concentrations of curcumin for 48 h and stained with Annexin-V. Apoptosis was analyzed by flow cytometry. (D) Curcumin inhibits the expression of anti-apoptotic proteins XIAP and survivin in HTLV-I-infected T-cell lines. Cells were treated with 50 μM curcumin for the indicated hours. Amounts of the anti-apoptotic proteins were determined by Western blotting.

Recently, we have shown that curcumin induces a strong anti-proliferative response in HTLV-I-infected leukemia cells [10]. Curcumin suppressed growth of HTLV-I-infected T-cell lines and peripheral blood mononuclear cells (PBMCs) from ATL patients in a dose-dependent fashion. In comparison, the growth inhibitory effect on PBMCs of healthy donors was weak (Figures 1A and 1B). These findings indicate that curcumin inhibits the growth of cells infected with HTLV-I but not uninfected PBMCs. The anti-proliferative effect of curcumin on HTLV-I-infected T cells was mediated through the induction of apoptosis. Curcumin induced a dose-dependent increase in proportion of cells positive for Annexin-V, which binds to the cells that express phosphatidylserine on the outer layer of the cell membrane, a characteristic feature of cells entering apoptosis, indicating increased apoptosis of curcumin-treated cells (Figure 1C). Rajasingh and colleagues [11] reported similar results, showing curcumin-induced apoptosis in HTLV-I-infected T-cell lines by DNA fragmentation and TUNEL assay. Curcumin-induced apoptosis is mediated by down-regulation of anti-apoptotic proteins, such as XIAP and survivin (Figure 1D). Although HTLV-I Tax protein plays a crucial role in leukemogenesis of HTLV-I-infected cells, the effect of curcumin on growth of HTLV-I-infected T-cell lines did not correlate with Tax expression level (Figure 1D). In addition, Tax is undetectable in circulating ATL cells by several mechanisms [12]. Curcumin inhibited growth of Tax-negative HTLV-I-infected T-cell lines, MT-1 and ED-40515(-), and primary ATL cells. Therefore, the growth inhibition of HTLV-I-infected T-cell lines and primary ATL cells induced by curcumin appears to be mediated by a Tax-independent pathway. We also examined the effect of curcumin on ATL *in vivo*. SCID mice, which were inoculated with HTLV-I-infected T-cell line HUT-102 cells, were treated with curcumin or vehicle orally once daily for 21 days. The mean tumor volume and weight of curcumin-treated mice were significantly lower than those of untreated mice (Figure 2). These results suggest that curcumin also has *in vivo* anti-ATL effect.

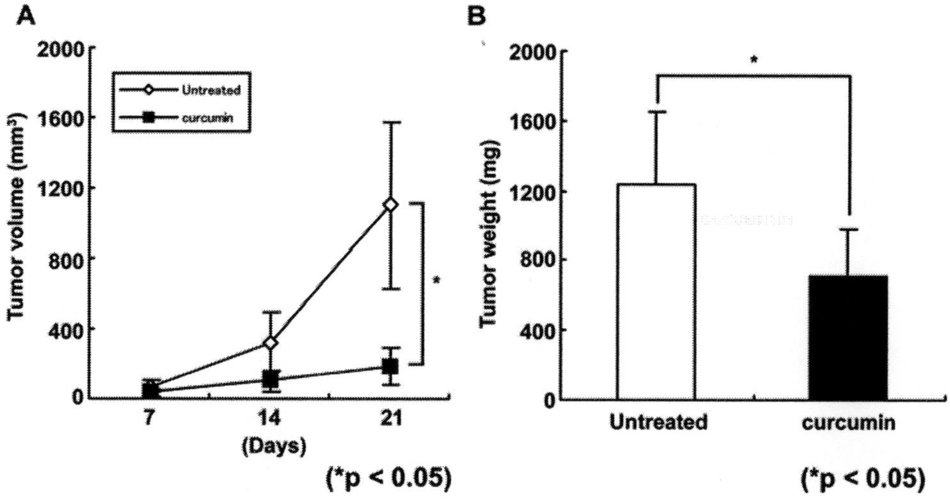

Figure 2. Effects of curcumin on growth of subcutaneous HTLV-I-infected T-cell line HUT-102 tumors in SCID mice. Mice were engrafted with 5×10^6 HUT-102 cells by subcutaneous injection in the post auricular region and were randomly placed into two cohorts of six mice each that received vehicle and curcumin treatment commenced on the day of injections. 300 mg/kg body weight curcumin was administered each day by gastric gavage for 21 consecutive days. Tumor size was measured three-dimensionally in the intact animal once a week using calipers. The tumors were excised on day 21 and weighted. (A) Serial changes in tumor volume in curcumin-treated and untreated mice. (B) Weight of tumors removed from curcumin-treated and untreated mice. Data are mean ± SD. *p <0.05, by Mann-Whitney U-test.

NF-κB PATHWAY

1. Regulation of NF-κB

NF-κB is a family of closely related protein dimers that bind to a common sequence motif in the DNA called the κB site [13]. NF-κB family consists of five structurally related proteins including c-Rel, RelA (p65), RelB, NF-κB1 (p50 and its precursor p105), and NF-κB2 (p52 and its precursor p100) [14]. NF-κB proteins share a highly conserved N-terminal Rel homology domain (RHD) responsible for DNA binding, dimerization, nuclear localization, and association with inhibitor of NF-κB (IκB) proteins that contain IκBα, IκBβ, IκBε, IκBγ, Bcl-3, and the precursor Rel proteins p100 and p105. NF-κB proteins can exist as homo- or heterodimers, and regulate the transcription of their target genes. In resting cells, the NF-κB proteins are sequestered in the cytoplasm through physical interaction with IκB proteins. The IκB proteins are characterized by the presence of multiple ankyrin repeats, which are protein-protein interaction domains that interact with NF-κB via the RHD. Cell stimulation activates IκB kinase (IKK) complex, which is composed of two catalytic subunits (IKKα/IKK1 and IKKβ/IKK2) and a regulatory subunit (IKKγ/NEMO) [15]. Activated IKK phosphorylates IκB proteins, and targets them for polyubiquitination by the SCFβ-TrCP ubiquitin ligase complex [16]. This modification subsequently targets IκB proteins for rapid degradation by the 26S proteasome. Freed NF-κB dimers translocate to the nucleus where they stimulate the transcriptional activity of several hundred target genes containing the consensus sequence 5'-GGGRNNYYCC-3'.

2. Anti-Apoptotic Effects of NF-κB in HTLV-I-Infected T Cells

Activation of NF-κB promotes survival and proliferation of various cancer cells. Down-regulation of NF-κB sensitizes cells to apoptosis through reduction of various anti-apoptotic gene products, such as inhibitor of apoptosis protein (IAP) family, Bcl-2 family, p53, and c-FLIP. HTLV-I Tax induces persistent activation of NF-κB [17]. Thus, NF-κB is constitutively present in the nuclei of Tax-expressing cells and HTLV-I-infected T cells. Through NF-κB, Tax constitutively activates the expression of various apoptosis regulatory genes such as cIAP-2 [18] XIAP [19], Bcl-x_L [20,21], and survivin [22]. Suppression of NF-κB in Tax-induced murine tumor cells sensitizes them to proapoptotic stimuli [23]. Leukemic cells from ATL patients express extremely low levels of viral genes including *tax* [12,24]. However, several genes that can be transactivated by Tax through NF-κB are already constitutively expressed in leukemic cells of ATL patients, probably by a Tax-independent mechanism [25]. Bay 11-7082, an inhibitor of NF-κB through irreversible inhibition of IκBα phosphorylation, induced apoptosis in HTLV-I-infected T-cell lines and primary ATL cells [26]. Recently, attempts to induce apoptosis of these cells by other NF-κB inhibitors were also successful [27-29]. These results indicate that NF-κB plays a crucial role in survival of HTLV-I-infected leukemia cells, and could be considered as a therapeutic target in ATL.

3. Effects of Curcumin on NF-κB Pathway in HTLV-I-Infected T Cells

We have examined the effects of curcumin on NF-κB activity in HTLV-I-infected T-cell lines and primary ATL cells [10]. Curcumin reduced constitutively activated NF-κB-DNA binding activity in HTLV-I-infected T-cell lines and primary ATL cells in a time-dependent fashion (Figures 3A and 3B). Curcumin also efficiently inhibited constitutively activated NF-κB transcriptional activity in an HTLV-I-infected T-cell line (Figure 3C, right panel). The expression of anti-apoptotic genes, such as XIAP and survivin, which were down-regulated by curcumin in a dose-dependent fashion (Figure 1D), is known to be regulated by NF-κB [19,22]. It is possible that inhibition of NF-κB activity by curcumin could induce apoptosis through down-regulation of these anti-apoptotic genes. Although Tax is an activator of NF-κB and Tax-induced NF-κB transcriptional activity in Jurkat cells is also inhibited by curcumin in a dose-dependent fashion (Figure 3C, left panel), curcumin-induced apoptosis is not associated with reduced expression of Tax (Figure 1D). How does curcumin inhibit NF-κB activity in HTLV-I-infected cells? Curcumin reduces the phosphorylated IκBα content in HTLV-I-infected T-cell lines and primary ATL cells, resulting in accumulation of IκBα (Figure 4). Thus, inhibition of NF-κB activity by curcumin is mediated through stabilization of IκBα by blocking phosphorylation and degradation of IκBα. Previous study has shown that curcumin suppresses the IKK activity needed for NF-κB activation [6,30-32]. It should be determined whether curcumin can inhibit phosphorylation of IκBα by inhibiting IKK activity in HTLV-I-infected T cells.

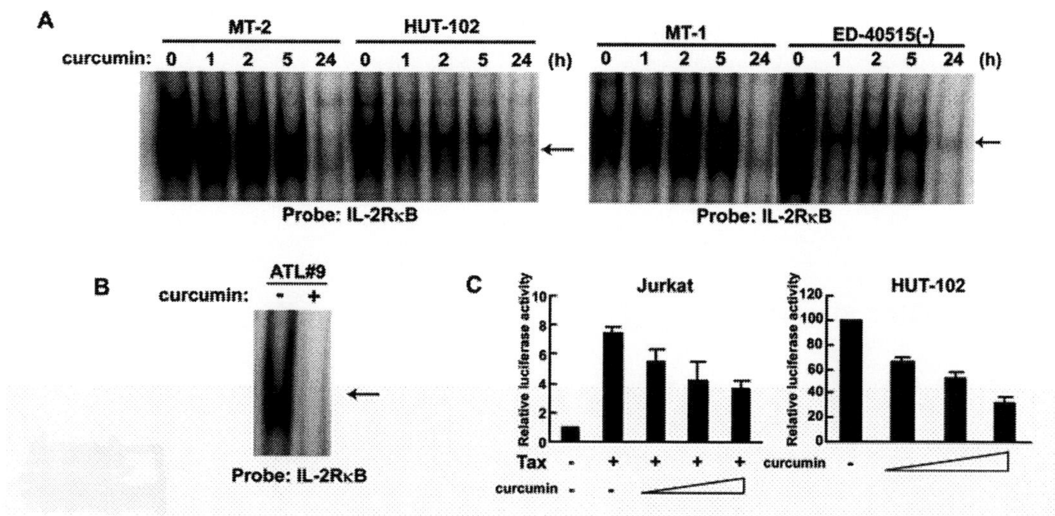

Figure 3. Curcumin inhibits NF-κB activity in HTLV-I-infected T-cell lines and primary ATL cells. Curcumin suppressed NF-κB-DNA-binding activity in HTLV-I-infected T-cell lines (A) and primary ATL cells (B). Nuclear extracts were incubated with a radiolabeled oligonucleotide that corresponds to the NF-κB sequence of the *IL-2Rα* gene and analyzed by gel-shift assay. (C) Curcumin inhibited both Tax-induced and constitutive NF-κB transcriptional activity. Reporter plasmid κB-LUC, luciferase expression plasmid controlled by five tandem repeats of the NF-κB binding sequence of the *IL2Rα* gene, was transfected into Jurkat cells with Tax-expression plasmid (left panel) or HUT-102 cells (right panel). After transfection, cells were treated with increasing concentrations of curcumin (0, 3.1, 12.5,

and 50 μM). Luciferase activity is expressed relative to the basal level measured in cells transfected with the reporter plasmid alone without further treatment, which was defined as 1 (left panel) and 100 (right panel). Data are mean ± SD of three independent experiments.

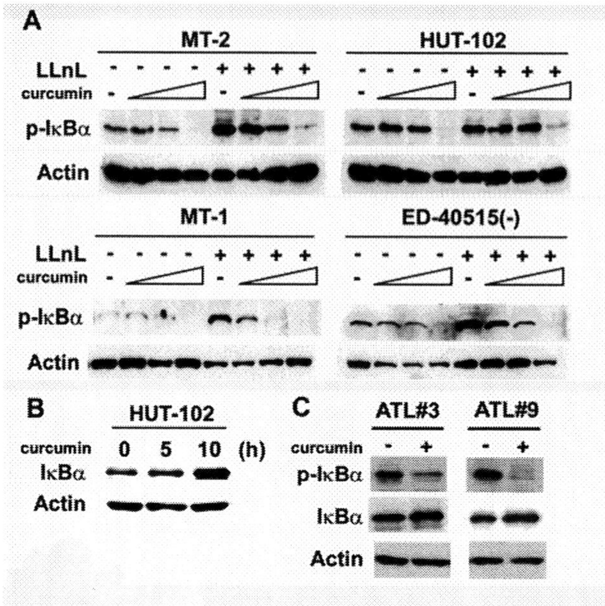

Figure 4. Curcumin inhibits phosphorylation of IκBα in HTLV-I-infected T-cell lines and primary ATL cells. (A) HTLV-I-infected T-cell lines were cultured in either presence (+) or absence (-) of proteasome inhibitor, LLnL (20 μM), for 1 h. Then cells were treated with increasing concentrations of curcumin (0, 3.1, 12.5, and 50 μM) for additional 4 h. (B) HUT-102 cells were treated with curcumin (50 μM) for the indicated time periods. (C) Primary ATL cells were treated with (+) or without (-) 50 μM curcumin for 24 h. Cell lysates were resolved on SDS-PAGE gels and analyzed by Western blotting with antibodies against phospho-IκBα (Ser32/36), total IκBα, and actin.

AP-1 Pathway

1. Regulation of AP-1

AP-1 was initially identified as a transcription factor that contributes to 12-*O*-tetradecanoyl-phorbol-13-acetate-inducible gene expression. Subsequent research identified AP-1 as a group of dimeric complexes formed by either homodimerization of two Jun (c-Jun, JunB, and JunD) families or heterodimerization of Jun and Fos (c-Fos, FosB, Fra-1, and Fra-2) families through the leucine zipper [33]. AP-1 recognizes a consensus element (5'-TGAG/CTCA-3') known as the AP-1 site present on the promoter regions of many genes induced by inflammation stimuli [34]. AP-1 can also act as an oncogenic transformation factor, and is required for proliferation of various human tumor cells [35].

2. Anti-Apoptotic Effects of AP-1 in HTLV-I-Infected T Cells

HTLV-I-infected T-cell lines express high levels of mRNA encoding c-Fos, Fra-1, c-Jun, JunB, and JunD, and exhibit high levels of AP-1 activity [36]. Tax can activate the expression of c-Fos, Fra-1, c-Jun, and JunD mRNAs [36]. Tax-induced activation of AP-1 is involved in the transcriptional activation of several cellular genes such as *IL-5*, *IL-8*, *TGF-β*, *fra-1*, *TR3/nur77*, *proenkephalin*, and *TIMP-1* [36]. Although expression of Tax is lacking, constitutive activation of AP-1 is observed in fresh primary uncultured ATL cells and in ATL-derived T-cell lines [37]. Activation seems to be certainly important in the development of ATL. Activated AP-1-DNA binding complexes in HTLV-I-infected T-cell lines and fresh ATL cells contained JunD [37]. Consistently, JunD mRNA is highly upregulated in these cells. The study of JunD$^{-/-}$ primary embryonic fibroblasts showed that these cells undergo premature growth arrest and displayed increased sensitivity to death-promoting agent, including UV irradiation or tumor necrosis factor-α [38]. Thus, constitutive activation of AP-1 containing JunD may inhibit apoptosis of ATL cells and could substantially contribute to their growth.

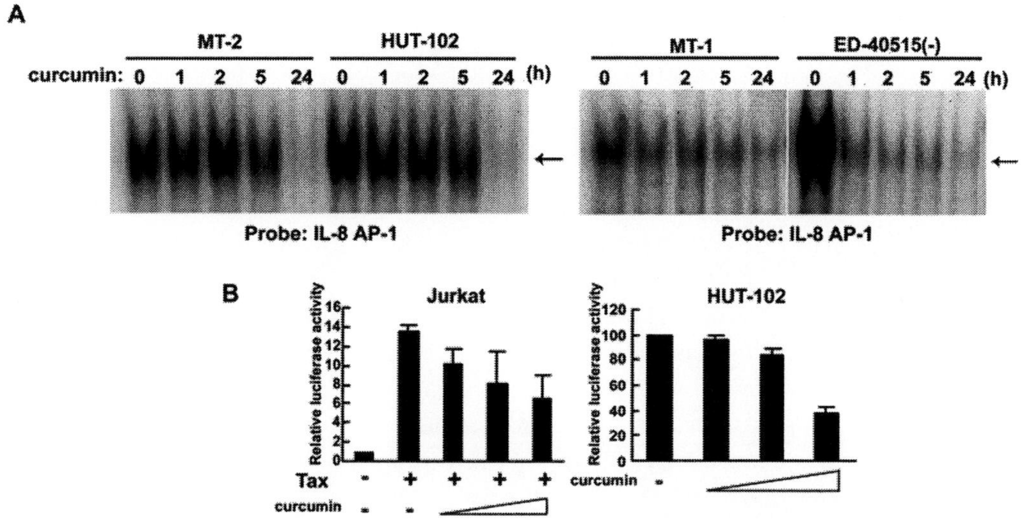

Figure 5. Curcumin inhibits AP-1 activity in HTLV-I-infected T-cell lines and primary ATL cells. (A) Curcumin (50 μM) suppressed AP-1-DNA-binding activity in HTLV-I-infected T-cell lines. Nuclear extracts were incubated with a radiolabeled oligonucleotide that corresponds to the AP-1 sequence of the *IL-8* gene and analyzed by gel-shift assay. (B) Curcumin inhibited both Tax-induced and constitutive AP-1 transcriptional activity. Reporter plasmid AP-1-LUC, luciferase expression plasmid controlled by two copies of the AP-1 binding sites of the *IL-8* gene, was transfected into Jurkat cells with Tax-expression plasmid (left panel) or HUT-102 cells (right panel). After transfection, cells were treated with increasing concentrations of curcumin (0, 3.1, 12.5, and 50 μM). Luciferase activity is expressed relative to the basal level measured in cells transfected with the reporter plasmid alone without further treatment, which was defined as 1 (left panel) and 100 (right panel). Data are mean ± SD of three independent experiments.

3. Effects of Curcumin on AP-1 Pathway in HTLV-I-Infected T Cells

We also examined the effects of curcumin on AP-1 activity in HTLV-I-infected T-cell lines [39]. Curcumin down-regulated constitutively activated AP-1-DNA binding activity in HTLV-I-infected T-cell lines in a time-dependent fashion (Figure 5A). It also inhibited Tax-induced and constitutively activated AP-1 transcriptional activity in a dose-dependent fashion (Figure 5B). Because increased AP-1-DNA binding complex in HTLV-I-infected T cells contained JunD (Figure 6A), we also examined the effect of curcumin on the expression of JunD. Curcumin reduced the expression of JunD protein in HTLV-I-infected T-cell lines in a time-dependent fashion (Figure 6B), suggesting that the inhibitory effect of curcumin on AP-1 activity is mediated by blocking the expression of JunD. Constitutive JNK activation was detected in HTLV-I-infected T cells [40]. JNK phosphorylation target sites are present in the JunD protein, and phosphorylation of JunD by JNK has been detected and phosphorylation of JunD by JNK can activate its transcriptional activity [41]. Curcumin has been reported to inhibit JNK activity [42]. Thus, the inhibitory effect of curcumin on AP-1 activity seems to be mediated by blocking the expression of JunD and JNK activity in HTLV-I-infected cells.

PI3K-Akt Pathway

1. Regulation of PI3K-Akt Pathway

Akt, a serine/threonine protein kinase that functions as a critical regulator of survival, proliferation and growth, is activated downstream of PI3K [43]. A variety of growth factors and other extracellular stimuli can activate PI3K through activation of their cognate receptor tyrosine kinases. The activated PI3K converts the plasma membrane lipid second messenger phosphatidylinositol-4,5-bisphosphate (PIP_2) to phosphatidylinositlo-3,4,5-triphosphate (PIP_3). Dephosphorylation of PIP_3 to regenerate PIP_2 is accomplished by the phosphatidylinositol 3-phosphatase PTEN. Akt is activated by recruitment to the plasma membrane through direct contact of its pleckstrin-homology domain with PIP_3, and phosphorylation at Thr308 and Ser473. Thr308 is phosphorylated by phosphoinositide-dependent kinase (PDK)1, whereas Ser473 is phosphorylated by PDK2 [44,45]. Activated Akt in turn regulates a wide range of target proteins known to regulate cell survival, proliferation, and growth. The mechanism by which Akt protects cells from apoptosis is likely to be multifactorial, because Akt directly phosphorylates several components of the apoptosis machinery. For example, Akt phosphorylates proapoptotic proteins, procaspase-9 [46] and Bad [47], resulting in blocking propagation of the apoptotic signaling. Phosphorylation of FoxO family of transcription factors, previously referred to as forkhead transcription factors, by Akt prevents its nuclear translocation and activation of FoxO gene targets (which include several proapoptotic genes, such as *Fas ligand*, *Bim*, and *TRADD*) [48]. PI3K-Akt signaling pathway can promote the activation of NF-κB and induce cell survival through promotion of anti-apoptotic gene expression [49]. Akt phosphorylates RelA through a p38 and IKKβ-dependent, or an IKKα-dependent mechanism. RelA phosphorylation enhances its transactivation potential by positively regulating RelA interactions with co-activators such as CBP and p300 [50]. Thus, activation of PI3K-Akt signaling pathway might play critical roles in anti-apoptosis in cancer cells. Indeed, aberrant deregulation of the PI3K-Akt pathway has been widely implicated in many human cancers, such as glioblastoma, ovarian cancer, breast cancer, endometrial cancer, hepatocellular carcinoma, melanoma, cancer in digestive tract, lung cancer, renal cell carcinoma, thyroid cancer, and lymphoma [51]. Manipulation of PI3K-Akt activity resulted in alteration of the response of tumor cells to chemotherapy and irradiation [52,53]. Therefore, compounds that block PI3K-Akt signaling pathway might be therapeutically useful in cancer treatment by blocking many aspects of the tumor cell phenotype.

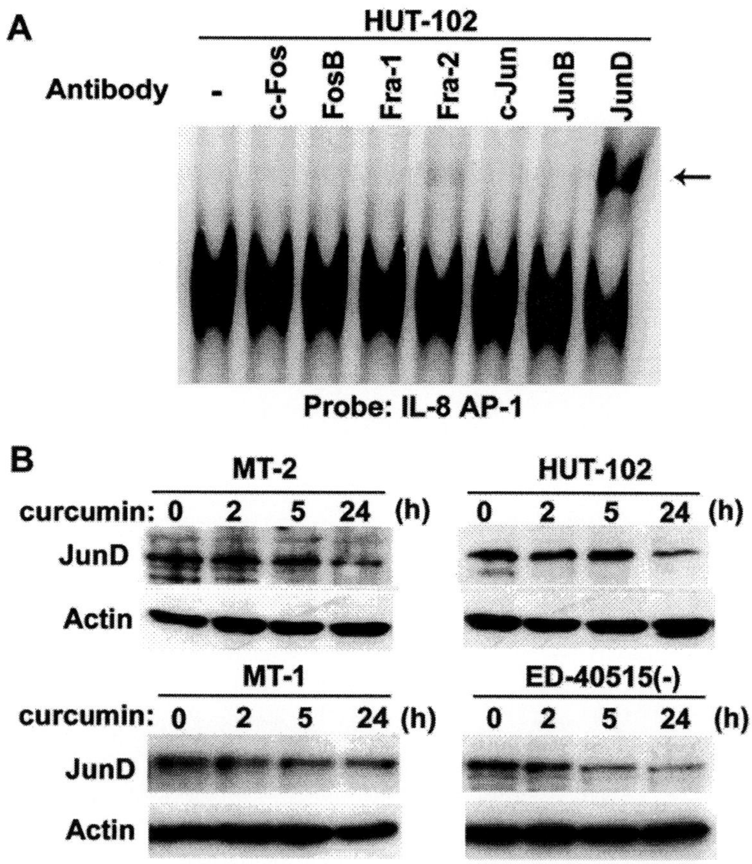

Figure 6. Curcumin inhibits the expression of JunD in HTLV-I-infected T-cell lines. (A) AP-1-DNA complexes contain JunD in an HTLV-I-infected T-cell line. AP-1 subunit specificity was determined by using nuclear extracts from untreated HUT-102 cells and antibodies to the AP-1 components. The super-shift band was observed following the addition of antibody against JunD (arrow). (B) HTLV-I-infected T-cell lines were treated with 50 μM curcumin for the indicated time period. Cell lysates were resolved on SDS-PAGE gels and analyzed by Western blotting with antibodies against JunD and actin.

2. Anti-Apoptotic Effects of PI3K-Akt in HTLV-I-Infected T Cells

Previous studies have shown that activation of PI3K-Akt signaling is involved in fibroblast Rat-1 transformation by HTLV-I Tax protein [54]. These findings suggest that Akt signaling pathway may be related to the cell transformation of HTLV-I-infected T cells, and that inhibitors of this pathway may be effective in the treatment of ATL. Akt is activated in HTLV-I-infected T-cell lines and primary ATL cells [55-58]. Tax activation of Akt is linked to NF-κB activation and p53 inhibition [57]. Furthermore, Tax also activates AP-1 through PI3K-Akt pathway [56]. In addition, we found that Tax could activate β-catenin through activation of Akt [59]. Inactivation of GSK-3β by Akt induced stabilization of β-catenin to activate β-catenin/Tcf-dependent transcription in Tax-positive HTLV-I-infected T-cell lines, resulting up-regulation of the genes associated with cell cycle progression and anti-apoptosis such as cyclin D1 and c-Myc [59]. The PI3K inhibitor LY294002 prevented growth of HTLV-I-infected T-cell lines and primary ATL cells by inducing apoptosis [55-57]. Moreover, overexpression of Akt dominant negative mutant inhibited growth of Tax-transformed cells [56]. These findings indicate that PI3K-Akt pathway is important for the survival of HTLV-I-infected T cells.

3. Effects of Curcumin on PI3K-Akt Pathway in HTLV-I-Infected T Cells

Recently, we have reported that curcumin inhibits PI3K-Akt pathway in HTLV-I-infected T-cell lines and primary ATL cells [55]. Curcumin reduced the constitutive phosphorylation of Akt at Ser473 in HTLV-I-infected T-cell lines and primary ATL cells (Figures 7A and 7B). Consistent with reduction of Akt phosphorylation, curcumin inhibited Akt kinase activity *in vitro* in a time-dependent fashion (Figure 7C). Phosphorylation of Akt at Thr308 by PDK1 is necessary and sufficient for Akt activation. Curcumin inhibited the activity of PDK1 by reducing phosphorylation of this protein (Figure 7D). Curcumin-induced inhibition of Akt resulted in reduction of phosphorylation of GSK-3β at Ser9, which is a downstream target of Akt (Figure 7D). Akt directly phosphorylates GSK-3β and blocks its kinase activity; thus, curcumin can stimulate GSK-3β activity. We also found that curcumin inhibited the expression of β-catenin protein through Akt inactivation in HTLV-I-infected T-cell lines (unpublished data).

JAK-STAT PATHWAY

1. Regulation of JAK-STAT Pathway

JAK-STAT constitutes an important signaling pathway that mediates cytokine-dependent growth and survival signals in immune cells. Binding of growth factors or cytokines to their receptors results in the activation of intrinsic receptor tyrosine kinase activity or of receptor-associated kinases including JAKs. JAKs subsequently phosphorylate the cytoplasmic tails of the receptor to provide docking sites for the recruitment of monomeric STATs. Once STATs have been recruited, they become substrates for tyrosine phosphorylation. In turn, phosphorylated STATs dimerize and translocate to the nucleus to activate the expression of genes necessary for cell proliferation and differentiation [60]. Four mammalian JAKs (JAK1, 2, 3, and TYK2) and seven STATs (STAT1, 2, 3, 4, 5a, 5b, and 6) family members have been identified. Constitutive activation of STAT1, 3, and 5 has been detected at high frequency in diverse human cancer cell lines and tissues, including leukemia, lymphoma, breast cancer, head and neck cancer, melanoma, ovarian cancer, lung cancer, pancreatic cancer, and prostate cancer [60]. These STATs-related signals promote the growth and survival of tumor cells, and thus contribute to malignancy.

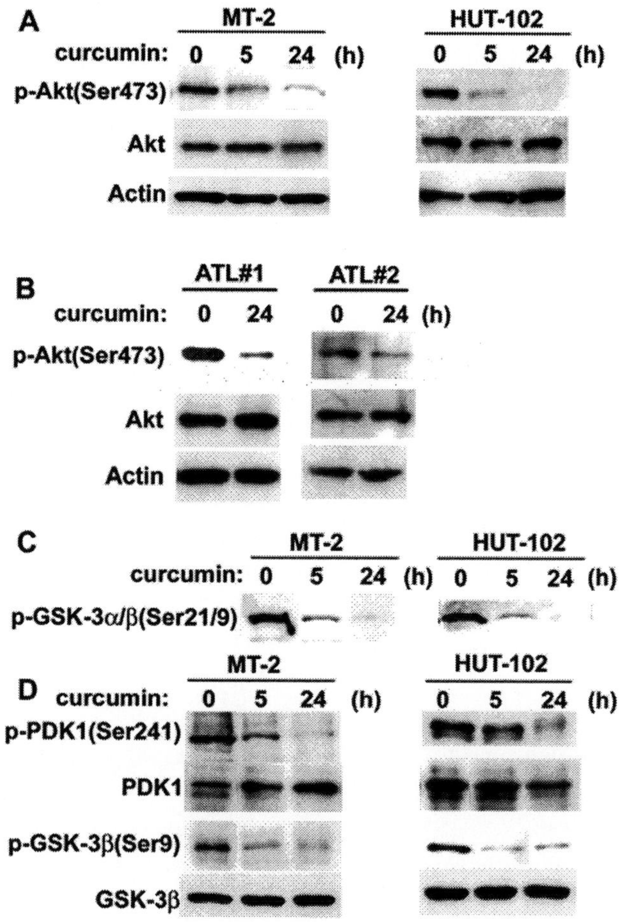

Figure 7. Curcumin inhibits constitutive activation of Akt in HTLV-I-infected T-cell lines and primary ATL cells. HTLV-I-infected T-cell lines (A) and primary ATL cells (B) were treated with 25 μM curcumin for indicated time periods. The phosphorylation status of Akt was assessed by Western blot analysis with anti-phospho-Akt (Ser473) and anti-Akt antibodies. (C) Effects of 25 μM curcumin on Akt kinase activity in HTLV-I-infected T-cell lines. *In vitro* Akt kinase assays with GSK-3α/β as a substrate were performed. Phosphorylated GSK-3α/β was detected with anti-GSK-3α/β (Ser21/9) antibody. (D) Effects of curcumin on phosphorylation status of proteins upstream and downstream of Akt. HTLV-I-infected T-cell lines were treated with 25 μM curcumin for indicated time periods. Phosphorylation of PDK1 and GSK-3β was assessed by Western blot analysis with anti-phospho-PDK1 (Ser241) and anti-phospho-GSK-3β (Ser9) antibodies.

2. Anti-Apoptotic Effects of JAK-STAT in HTLV-I-Infected T Cells

High expression of the interleukin-2 receptor α chain (IL-2Rα) is a common feature of ATL cells and HTLV-I-infected T-cell lines [61]. JAK-STAT is one of the well documented signaling pathways mediated by IL-2R [62]. Constitutive activation of the IL-2R-JAK-STAT signaling pathway correlates with IL-2 independence of HTLV-I-transformed T-cell lines [63]. While the JAK and STAT proteins usually remain unphosphorylated and inactive in quiescent lymphocytes, the HTLV-I-transformed T-cell lines express constitutively phosphorylated JAK1, JAK3, STAT3, and STAT5 proteins [63-66]. Similarly, an *in vitro* study using uncultured leukemic cells from HTLV-I seropositive patients with ATL also displayed constitutive activation of JAKs and STATs [67]. Although activation of JAK-STAT pathway seems to be important in growth and survival of HTLV-I-infected T cells, previous study showed that treatment with AG490, a JAK-specific inhibitor, suppressed STAT but had no effect on growth of HTLV-I-infected T-cell lines [68]. In contrast, our laboratory has demonstrated that AG490 inhibited the growth of HTLV-I-infected T-cell lines and primary ATL cells by inducing G_1 cell cycle arrest via the altered expression of cyclin D2, Cdk4, p53, p21, Pim-1, and c-Myc, and by inducing apoptosis by reducing the expression of cIAP-2, XIAP, survivin, and Bcl-2 [69]. These results indicate that activation of JAK-STAT pathway is responsible for the proliferation and survival of ATL cells.

3. Effects of Curcumin on JAK-STAT Pathway in HTLV-I-Infected T Cells

Recently, Rajasingh and colleagues [11] demonstrated that curcumin inhibited the constitutive phosphorylation of STAT3, STAT5, JAK3, and TYK2 in HTLV-I-infected T-cell lines. The exact mechanism of action of curcumin on JAK-STAT signaling pathway in HTLV-I-infected T-cell lines is not known. While the inhibition of STAT proteins may be due to inhibition of upstream JAK kinases, curcumin might directly act on JAK and STAT proteins. Furthermore, it is possible that curcumin activates the negative regulators of JAK-STAT pathway such as suppressors of cytokine signaling (SOCS) or Src homology phosphatase (SHP), which in turn modulates the constitutive activation of JAK-STAT pathway in leukemia [70,71]. A similar mechanism was described in brain microglial cells where curcumin suppressed JAK-STAT signaling by activating SHP-2, and thus attenuated the inflammatory response of these cells [72].

CONCLUSION

Figure 8 provides a schematic diagram of the various mechanisms of apoptosis and cell cycle arrest induced by curcumin. Because curcumin inhibits various signaling pathways such as NF-κB, AP-1, PI3K-Akt, and JAK-STAT, which are associated with leukemogenesis of ATL, we propose that curcumin is a potentially promising compound for the treatment of ATL. Curcumin hardly inhibited the survival of normal PBMCs and systemic treatment of curcumin caused no obvious toxicity in mice [10]. Moreover, phase I clinical trials indicate that curcumin is tolerated in extremely large oral doses without apparent toxicity in human [9]. These results suggest that curcumin is a pharmacologically safe agent. Therefore, curcumin is a potentially useful anti-ATL drug. Most modern medicines currently available for treatment of ATL are very expensive, toxic, and less effective. Thus, one must investigate in more detail the agents that can be derived from natural sources, described traditionally, for the prevention and treatment of ATL. More clinical trials are also needed to validate the usefulness of curcumin either alone or in combination with existing therapy.

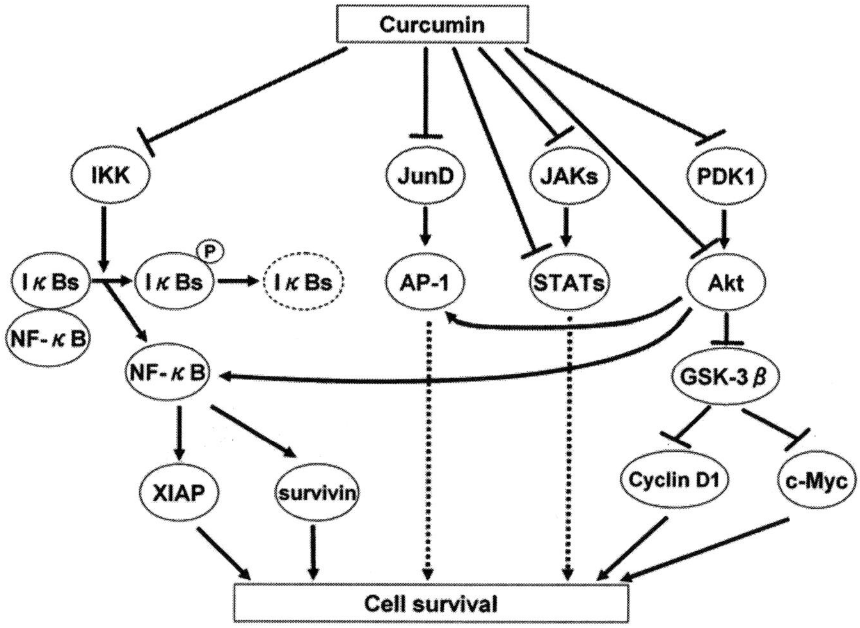

Figure 8. Schematic diagram of possible mechanisms of apoptosis and cell cycle arrest induced by curcumin.

References

- Yamada, Y., Tomonaga, M., Fukuda, H., Hanada, S., Utsunomiya, A., Tara, M., Sano, M., Ikeda, S., Takatsuki, K., Kozuru, M., Araki, K., Kawano, F., Niimi, M., Tobinai, K., Hotta, T. and Shimoyama, M. (2001). A new G-CSF-supported combination chemotherapy, LSG15, for adult T-cell leukaemia-lymphoma: Japan Clinical Oncology Group Study 9303. *Br. J. Haematol.*, *113*, 375-382.
- Grassmann, R., Aboud, M. and Jeang, K. T. (2005). Molecular mechanisms of cellular transformation by HTLV-1 Tax. *Oncogene*, *24*, 5976-5985.
- Sharma, R. A., Gescher, A. J. and Steward, W. P. (2005). Curcumin: The story so far. *Eur. J. Cancer*, *41*, 1955-1968.
- Anto, R. J., Mukhopadhyay, A., Denning, K. and Aggarwal, B. B. (2002). Curcumin (diferuloylmethane) induces apoptosis through activation of caspase-8,

BID cleavage and cytochrome c release: its suppression by ectopic expression of Bcl-2 and Bcl-xl. *Carcinogenesis*, *23*, 143-150.

- Shishodia, S., Amin, H. M., Lai, R. and Aggarwal, B. B. (2005). Curcumin (diferuloylmethane) inhibits constitutive NF-κB activation, induces G1/S arrest, suppresses proliferation, and induces apoptosis in mantle cell lymphoma. *Biochem. Pharmacol.*, *70*, 700-713.
- Bharti, A. C., Donato, N., Singh, S. and Aggarwal, B. B. (2003). Curcumin (diferuloylmethane) down-regulates the constitutive activation of nuclear factor-κB and IκBα kinase in human multiple myeloma cells, leading to suppression of proliferation and induction of apoptosis. *Blood*, *101*, 1053-1062.
- Sharma, R. A., Ireson, C. R., Verschoyle, R. D., Hill, K. A., Williams, M. L., Leuratti, C., Manson, M. M., Marnett, L. J., Steward, W. P. and Gescher, A. (2001). Effects of dietary curcumin on glutathione S-transferase and malondialdehyde-DNA adducts in rat liver and colon mucosa: relationship with drug levels. *Clin. Cancer Res.*, *7*, 1452-1458.
- Perkins, S., Verschoyle, R. D., Hill, K., Parveen, I., Threadgill, M. D., Sharma, R. A., Williams, M. L., Steward, W. P. and Gescher, A. J. (2002). Chemopreventive efficacy and pharmacokinetics of curcumin in the min/+ mouse, a model of familial adenomatous polyposis. *Cancer Epidemiol. Biomarkers Prev.*, *11*, 535-540.
- Sharma, R. A., Euden, S. A., Platton, S. L., Cooke, D. N., Shafayat, A., Hewitt, H. R., Marczylo, T. H., Morgan, B., Hemingway, D., Plummer, S. M., Pirmohamed, M., Gescher, A. J. and Steward, W. P. (2004). Phase I clinical trial of oral curcumin: biomarkers of systemic activity and compliance. *Clin. Cancer Res.*, *10,* 6847-6854.
- Tomita, M., Kawakami, H., Uchihara, J. N., Okudaira, T., Masuda, M., Takasu, N., Matsuda, T., Ohta, T., Tanaka, Y., Ohshiro, K. and Mori, N. (2006). Curcumin (diferuloylmethane) inhibits constitutive active NF-κB, leading to suppression of cell growth of human T-cell leukemia virus type I-infected T-cell lines and primary adult T-cell leukemia cells. *Int. J. Cancer*, *118*, 765-772.
- Rajasingh, J., Raikwar, H. P., Muthian, G., Johnson, C. and Bright, J. J. (2006). Curcumin induces growth-arrest and apoptosis in association with the inhibition of constitutively active JAK-STAT pathway in T cell leukemia. *Biochem. Biophys. Res. Commun.*, *340*, 359-368.
- Matsuoka, M. (2005). Human T-cell leukemia virus type I (HTLV-I) infection and the onset of adult T-cell leukemia (ATL). *Retrovirology*, *2*, 27.
- Garg, A. and Aggarwal, B. B. (2002). Nuclear transcription factor-κB as a target for cancer drug development. *Leukemia*, *16*, 1053-1068.
- Ghosh, S. and Karin, M. (2002). Missing pieces in the NF-κB puzzle. *Cell*, *109* Suppl, S81-96.
- Rothwarf, D. M. and Karin, M. (1999). The NF-κB activation pathway: a paradigm in information transfer from membrane to nucleus. *Sci. STKE*, 1999, RE1.
- Karin, M. and Ben-Neriah, Y. (2000). Phosphorylation meets ubiquitination: the control of NF-κB activity. *Annu. Rev. Immunol.*, *18*, 621-663.
- Sun, S. C. and Yamaoka, S. (2005). Activation of NF-κB by HTLV-I and implications for cell transformation. *Oncogene*, *24*, 5952-5964.

- Wäldele, K., Silbermann, K., Schneider, G., Ruckes, T., Cullen, B. R. and Grassmann, R. (2006). Requirement of the human T-cell leukemia virus (HTLV-1) tax-stimulated HIAP-1 gene for the survival of transformed lymphocytes. *Blood, 107,* 4491-4499.
- Kawakami, A., Nakashima, T., Sakai, H., Urayama, S., Yamasaki, S., Hida, A., Tsuboi, M., Nakamura, H., Ida, H., Migita, K., Kawabe, Y. and Eguchi, K. (1999). Inhibition of caspase cascade by HTLV-I tax through induction of NF-κB nuclear translocation. *Blood, 94,* 3847-3854.
- Tsukahara, T., Kannagi, M., Ohashi, T., Kato, H., Arai, M., Nunez, G., Iwanaga, Y., Yamamoto, N., Ohtani, K., Nakamura, M. and Fujii, M. (1999). Induction of Bcl-x(L) expression by human T-cell leukemia virus type 1 Tax through NF-κB in apoptosis-resistant T-cell transfectants with Tax. *J. Virol., 73,* 7981-7987.
- Mori, N., Fujii, M., Cheng, G., Ikeda, S., Yamasaki, Y., Yamada, Y., Tomonaga, M. and Yamamoto, N. (2001). Human T-cell leukemia virus type I tax protein induces the expression of anti-apoptotic gene Bcl-xL in human T-cells through nuclear factor-κB and c-AMP responsive element binding protein pathways. *Virus Genes, 22,* 279-287.
- Kawakami, H., Tomita, M., Matsuda, T., Ohta, T., Tanaka, Y., Fujii, M., Hatano, M., Tokuhisa, T. and Mori, N. (2005). Transcriptional activation of survivin through the NF-κB pathway by human T-cell leukemia virus type I tax. *Int. J. Cancer, 115,* 967-974.
- Portis, T., Harding, J. C. and Ratner, L. (2001). The contribution of NF-κB activity to spontaneous proliferation and resistance to apoptosis in human T-cell leukemia virus type 1 Tax-induced tumors. *Blood, 98,* 1200-1208.
- Kinoshita, T., Shimoyama, M., Tobinai, K., Ito, M., Ito, S., Ikeda, S., Tajima, K., Shimotohno, K. and Sugimura, T. (1989). Detection of mRNA for the tax1/rex1 gene of human T-cell leukemia virus type I in fresh peripheral blood mononuclear cells of adult T-cell leukemia patients and viral carriers by using the polymerase chain reaction. *Proc. Natl. Acad. Sci. U S A, 86,* 5620-5624.
- Mori, N., Fujii, M., Ikeda, S., Yamada, Y., Tomonaga, M., Ballard, D. W. and Yamamoto, N. (1999). Constitutive activation of NF-κB in primary adult T-cell leukemia cells. *Blood, 93,* 2360-2368.
- Mori, N., Yamada, Y., Ikeda, S., Yamasaki, Y., Tsukasaki, K., Tanaka, Y., Tomonaga, M., Yamamoto, N. and Fujii, M. (2002). Bay 11-7082 inhibits transcription factor NF-κB and induces apoptosis of HTLV-I-infected T-cell lines and primary adult T-cell leukemia cells. *Blood, 100,* 1828-1834.
- Satou, Y., Nosaka, K., Koya, Y., Yasunaga, J. I., Toyokuni, S. and Matsuoka, M. (2004). Proteasome inhibitor, bortezomib, potently inhibits the growth of adult T-cell leukemia cells both in vivo and in vitro. *Leukemia, 18,* 1357-1363.
- Watanabe, M., Ohsugi, T., Shoda, M., Ishida, T., Aizawa, S., Maruyama-Nagai, M., Utsunomiya, A., Koga, S., Yamada, Y., Kamihira, S., Okayama, A., Kikuchi, H., Uozumi, K., Yamaguchi, K., Higashihara, M., Umezawa, K., Watanabe, T. and Horie, R. (2005). Dual targeting of transformed and untransformed HTLV-1-

infected T cells by DHMEQ, a potent and selective inhibitor of NF-κB, as a strategy for chemoprevention and therapy of adult T-cell leukemia. *Blood, 106*, 2462-2471.

- Sanda, T., Asamitsu, K., Ogura, H., Iida, S., Utsunomiya, A., Ueda, R. and Okamoto, T. (2006). Induction of cell death in adult T-cell leukemia cells by a novel IκB kinase inhibitor. *Leukemia, 20*, 590-598.
- Plummer, S. M., Holloway, K. A., Manson, M. M., Munks, R. J., Kaptein, A., Farrow, S. and Howells, L. (1999). Inhibition of cyclo-oxygenase 2 expression in colon cells by the chemopreventive agent curcumin involves inhibition of NF-κB activation via the NIK/IKK signalling complex. *Oncogene, 18*, 6013-6020.
- Jobin, C., Bradham, C. A., Russo, M. P., Juma, B., Narula, A. S., Brenner, D. A. and Sartor, R. B. (1999). Curcumin blocks cytokine-mediated NF-κB activation and proinflammatory gene expression by inhibiting inhibitory factor I-κB kinase activity. *J. Immunol., 163*, 3474-3483.
- Pan, M. H., Lin-Shiau, S. Y. and Lin, J. K. (2000). Comparative studies on the suppression of nitric oxide synthase by curcumin and its hydrogenated metabolites through down-regulation of IκB kinase and NFκB activation in macrophages. *Biochem. Pharmacol., 60*, 1665-1676.
- Curran, T. and Franza, B. R., Jr. (1988). Fos and Jun: the AP-1 connection. *Cell, 55*, 395-397.
- Chinenov, Y. and Kerppola, T. K. (2001). Close encounters of many kinds: Fos-Jun interactions that mediate transcription regulatory specificity. *Oncogene, 20,* 2438-2452.
- Shaulian, E. and Karin, M. (2002). AP-1 as a regulator of cell life and death. *Nat. Cell Biol., 4*, E131-136.
- Hall, W. W. and Fujii, M. (2005). Deregulation of cell-signaling pathways in HTLV-1 infection. *Oncogene, 24,* 5965-5975.
- Mori, N., Fujii, M., Iwai, K., Ikeda, S., Yamasaki, Y., Hata, T., Yamada, Y., Tanaka, Y., Tomonaga, M. and Yamamoto, N. (2000). Constitutive activation of transcription factor AP-1 in primary adult T- cell leukemia cells. *Blood, 95*, 3915-3921.
- Weitzman, J. B., Fiette, L., Matsuo, K. and Yaniv, M. (2000). JunD protects cells from p53-dependent senescence and apoptosis. *Mol. Cell, 6*, 1109-1119.
- Tomita, M., Kawakami, H., Uchihara, J. N., Okudaira, T., Masuda, M., Takasu, N., Matsuda, T., Ohta, T., Tanaka, Y. and Mori, N. (2006). Curcumin suppresses constitutive activation of AP-1 by downregulation of JunD protein in HTLV-1-infected T-cell lines. *Leuk. Res., 30*, 313-321.
- Xu, X., Heidenreich, O., Kitajima, I., McGuire, K., Li, Q., Su, B. and Nerenberg, M. (1996). Constitutively activated JNK is associated with HTLV-1 mediated tumorigenesis. *Oncogene, 13*, 135-142.
- Yazgan, O. and Pfarr, C. M. (2002). Regulation of two JunD isoforms by Jun N-terminal kinases. *J. Biol. Chem., 277*, 29710-29718.
- Chen, Y. R. and Tan, T. H. (1998). Inhibition of the c-Jun N-terminal kinase (JNK) signaling pathway by curcumin. *Oncogene, 17*, 173-178.

- Vivanco, I. and Sawyers, C. L. (2002). The phosphatidylinositol 3-Kinase AKT pathway in human cancer. *Nat. Rev. Cancer*, *2*, 489-501.
- Stephens, L., Anderson, K., Stokoe, D., Erdjument-Bromage, H., Painter, G. F., Holmes, A. B., Gaffney, P. R., Reese, C. B., McCormick, F., Tempst, P., Coadwell, J. and Hawkins, P. T. (1998). Protein kinase B kinases that mediate phosphatidylinositol 3,4,5-trisphosphate-dependent activation of protein kinase B. *Science*, *279*, 710-714.
- Toker, A. and Newton, A. C. (2000). Akt/protein kinase B is regulated by autophosphorylation at the hypothetical PDK-2 site. *J. Biol. Chem.*, *275*, 8271-8274.
- Cardone, M. H., Roy, N., Stennicke, H. R., Salvesen, G. S., Franke, T. F., Stanbridge, E., Frisch, S. and Reed, J. C. (1998). Regulation of cell death protease caspase-9 by phosphorylation. *Science*, *282*, 1318-1321.
- Datta, S. R., Dudek, H., Tao, X., Masters, S., Fu, H., Gotoh, Y. and Greenberg, M. E. (1997). Akt phosphorylation of BAD couples survival signals to the cell-intrinsic death machinery. *Cell*, *91*, 231-241.
- Datta, S. R., Brunet, A. and Greenberg, M. E. (1999). Cellular survival: a play in three Akts. *Genes Dev.*, *13*, 2905-2927.
- Ozes, O. N., Mayo, L. D., Gustin, J. A., Pfeffer, S. R., Pfeffer, L. M. and Donner, D. B. (1999). NF-κB activation by tumour necrosis factor requires the Akt serine-threonine kinase. *Nature*, *401*, 82-85.
- Viatour, P., Merville, M. P., Bours, V. and Chariot, A. (2005). Phosphorylation of NF-κB and IκB proteins: implications in cancer and inflammation. *Trends. Biochem. Sci.*, *30*, 43-52.
- Luo, J., Manning, B. D. and Cantley, L. C. (2003). Targeting the PI3K-Akt pathway in human cancer: rationale and promise. *Cancer Cell*, *4*, 257-262.
- Krystal, G. W., Sulanke, G. and Litz, J. (2002). Inhibition of phosphatidylinositol 3-kinase-Akt signaling blocks growth, promotes apoptosis, and enhances sensitivity of small cell lung cancer cells to chemotherapy. *Mol. Cancer Ther.*, *1*, 913-922.
- Brognard, J., Clark, A. S., Ni, Y. and Dennis, P. A. (2001). Akt/protein kinase B is constitutively active in non-small cell lung cancer cells and promotes cellular survival and resistance to chemotherapy and radiation. *Cancer Res.*, *61*, 3986-3997.
- Liu, Y., Wang, Y., Yamakuchi, M., Masuda, S., Tokioka, T., Yamaoka, S., Maruyama, I. and Kitajima, I. (2001). Phosphoinositide-3 kinase-PKB/Akt pathway activation is involved in fibroblast Rat-1 transformation by human T-cell leukemia virus type I tax. *Oncogene*, *20*, 2514-2526.
- Tomita, M., Matsuda, T., Kawakami, H., Uchihara, J. N., Okudaira, T., Masuda, M., Ohshiro, K. and Mori, N. (2006). Curcumin targets Akt cell survival signaling pathway in HTLV-I-infected T-cell lines. *Cancer Sci.*, *97*, 322-327.
- Peloponese, J. M., Jr. and Jeang, K. T. (2006). Role for Akt/protein kinase B and activator protein-1 in cellular proliferation induced by the human T-cell leukemia virus type 1 tax oncoprotein. *J. Biol. Chem.*, *281*, 8927-8938.

- Jeong, S. J., Pise-Masison, C. A., Radonovich, M. F., Park, H. U. and Brady, J. N. (2005). Activated AKT regulates NF-κB activation, p53 inhibition and cell survival in HTLV-1-transformed cells. *Oncogene*, *24*, 6719-6728.
- Fukuda, R., Hayashi, A., Utsunomiya, A., Nukada, Y., Fukui, R., Itoh, K., Tezuka, K., Ohashi, K., Mizuno, K., Sakamoto, M., Hamanoue, M. and Tsuji, T. (2005). Alteration of phosphatidylinositol 3-kinase cascade in the multilobulated nuclear formation of adult T cell leukemia/lymphoma (ATLL). *Proc. Natl. Acad. Sci. U S A*, *102*, 15213-15218.
- Tomita, M., Kikuchi, A., Akiyama, T., Tanaka, Y. and Mori, N. (2006). Human T-cell leukemia virus type 1 Tax dysregulates β-catenin signaling. *J. Virol.*, 80, 10497-10505.
- Yu, H. and Jove, R. (2004). The Stats of cancer-new molecular targets come of age. *Nat. Rev. Cancer*, *4*, 97-105.
- Franchini, G. (1995). Molecular mechanisms of human T-cell leukemia/lymphotropic virus type I infection. *Blood*, *86*, 3619-3639.
- Schindler, C. and Darnell, J. E., Jr. (1995). Transcriptional responses to polypeptide ligands: the JAK-STAT pathway. *Annu. Rev. Biochem.*, *64,* 621-651.
- Migone, T. S., Lin, J. X., Cereseto, A., Mulloy, J. C., O'Shea, J. J., Franchini, G. and Leonard, W. J. (1995). Constitutively activated Jak-STAT pathway in T cells transformed with HTLV-I. *Science*, *269*, 79-81.
- Cheng, J., Zhang, D., Zhou, C. and Marasco, W. A. (2004). Down-regulation of SHP1 and up-regulation of negative regulators of JAK/STAT signaling in HTLV-1 transformed cell lines and freshly transformed human peripheral blood CD4+ T-cells. *Leuk. Res.*, *28*, 71-82.
- Nakamura, N., Fujii, M., Tsukahara, T., Arai, M., Ohashi, T., Wakao, H., Kannagi, M. and Yamamoto, N. (1999). Human T-cell leukemia virus type 1 Tax protein induces the expression of *STAT1* and *STAT5* genes in T-cells. *Oncogene*, *18*, 2667-2675.
- Zhang, Q., Lee, B., Korecka, M., Li, G., Weyland, C., Eck, S., Gessain, A., Arima, N., Lessin, S. R., Shaw, L. M., Luger, S., Kamoun, M. and Wasik, M. A. (1999). Differences in phosphorylation of the IL-2R associated JAK/STAT proteins between HTLV-I(+), IL-2-independent and IL-2-dependent cell lines and uncultured leukemic cells from patients with adult T-cell lymphoma/leukemia. *Leuk. Res.*, *23*, 373-384.
- Takemoto, S., Mulloy, J. C., Cereseto, A., Migone, T. S., Patel, B. K., Matsuoka, M., Yamaguchi, K., Takatsuki, K., Kamihira, S., White, J. D., Leonard, W. J., Waldmann, T. and Franchini, G. (1997). Proliferation of adult T cell leukemia/lymphoma cells is associated with the constitutive activation of JAK/STAT proteins. *Proc. Natl. Acad. Sci. USA*, *94*, 13897-13902.
- Kirken, R. A., Erwin, R. A., Wang, L., Wang, Y., Rui, H. and Farrar, W. L. (2000). Functional uncoupling of the Janus kinase 3-Stat5 pathway in malignant growth of human T cell leukemia virus type 1-transformed human T cells. *J. Immunol.*, *165*, 5097-5104.

- Tomita, M., Kawakami, H., Uchihara, J. N., Okudaira, T., Masuda, M., Matsuda, T., Tanaka, Y., Ohshiro, K. and Mori, N. (2006). Inhibition of constitutively active Jak-Stat pathway suppresses cell growth of human T-cell leukemia virus type 1-infected T-cell lines and primary adult T-cell leukemia cells. *Retrovirology*, *3*, 22.
- Alexander, W. S. and Hilton, D. J. (2004). The role of suppressors of cytokine signaling (SOCS) proteins in regulation of the immune response. *Annu. Rev. Immunol.*, *22*, 503-529.
- Kile, B. T., Nicola, N. A. and Alexander, W. S. (2001). Negative regulators of cytokine signaling. *Int. J. Hematol.*, *73*, 292-298.
- Kim, H. Y., Park, E. J., Joe, E. H. and Jou, I. (2003). Curcumin suppresses Janus kinase-STAT inflammatory signaling through activation of Src homology 2 domain-containing tyrosine phosphatase 2 in brain microglia. *J. Immunol.*, *171*, 6072-6079.

In: Cell Apoptosis and Cancer
Editor: Albina W. Taylor, pp. 135-153
ISBN: 978-1-60021-506-3

Chapter VIII

HYPERTHERMIA-INDUCED APOPTOSIS

***E.H.A. Nijhuis*[1]*, A.A. Poot*[1]*, J. Feijen*[1] *and I. Vermes*[1,2]**
[1]University of Twente, Faculty of Science and Technology, Polymer Chemistry and Biomaterials Group, P.O. Box 217, 7500 AE Enschede, The Netherlands;
[2]Department of Clinical Chemistry, Hospital Group Medisch Spectrum Twente, P.O. Box 50.000, 7500 KA Enschede, The Netherlands.

1. INTRODUCTION

The idea of treating cancer cells by means of hyperthermia dates back at least 5000 years. According to the Edwin Smith surgical papyrus a fire drill was used by "surgeons" to cauterize carcinoma in the breast. Over the past 30 years there are significant improvements in this "war against cancer". Despite this, the mortality of patients receiving treatments that die of or with their cancer is relatively high. The failure to achieve a reduction in mortality from some forms of cancer using conventional treatments, namely surgery, radiotherapy and chemotherapy has added a new dimension of interest in the potential of hyperthermia, both as treatment in its own right and in combination with other therapies.

The main reason why there has been so little interest for hyperthermia in modern cancer research is that the technologies used in the past cannot deliver effective and homogeneous heating of all sites, particularly the deeper seeded tumors. Also research and implementation has been hampered by lack of non-invasive temperature measurement. Recently, development of new techniques (nanotechnology, computer modelling, non-invasive thermometry) to control and direct the heat, stimulated the interest in hyperthermia again.

The objective of this artificially induced temperature rise is the treatment of tumors, directly by introducing irreversible biological damage or indirectly by enhancing the effects of other treatment regimes such as X-irradiation or chemotherapy in a synergistic way. Although the ultimate aim in cancer therapy is to kill cancer cells, the way the cells die is of importance. Nature employs two ways to kill cells; namely by necrosis and apoptosis. Necrosis usually induces an undesirable inflammatory response in vivo, while apoptotic death of cells is relatively free from this post-treatment problem. Apoptosis, characterized as programmed cell death, is considered a natural process in the homeostasis and defence systems of the body. Hyperthermia could be used in cancer therapy to induce apoptosis instead of necrosis to minimize adverse side effects. Although hyperthermia has been under investigation for decades, the precise mechanisms how hyperthermia induces apoptosis remains unclear. In order to further increase the effectiveness of hyperthermia the cellular mechanisms must be elucidated. Here, we outline the effects of hyperthermia both on cellular and molecular levels. Furthermore, several ways to induce hyperthermia and novel therapeutic strategies will be discussed.

2. Physiological Features of Hyperthermia

The cell killing property of hyperthermia and its non-invasive character makes it an interesting technique to further exploit in combination with other cancer therapies or perhaps as a therapy alone. Oversimplified, the amount of cell killing depends on the thermal dose, which is a function of temperature and time. However, due to the complexity of heat-induced changes in cells and tissues, the temperature needed and the thermal dose applied for hyperthermia is still unresolved and heavily discussed. Although a large number of investigations concerning the cellular effects on hyperthermia exist, dating back to the 1970s and early-80s, the fundamental aspects of this technique is far from being elucidated. Here, several cellular effectors of hyperthermia are being described.

2.1. Cell Cycle

Synchronized cell cultures exhibit variations in their susceptibility to heat in accordance to their phase in the cell cycle. In general, the highest heat sensitivity can be observed during the mitotic phase. Microscopic examinations of M-phase cells exposed to hyperthermia show damage of their mitotic apparatus leading to insufficient mitosis. S-phase cells are also sensitive to hyperthermia, where chromosomal damage is observed. Both S- and M-phase cells undergo a ‘slow mode of cell death’ after hyperthermia, whereas those exposed to heat during G1-phase are relatively heat resistant and do not show any microscopic damage. These variations existing between the different cell cycle phases indicate the possible diversity of molecular mechanisms of cell death following hyperthermia [1].

2.2. Cell Membrane

Hyperthermia affects fluidity and stability of cellular membranes and impedes the function of transmembranal transport proteins and cell surface receptors in vitro. These findings suggested that membrane alterations represent an important target in hyperthermic cell death. These observations give raise to numerous reports on change in membrane potential, elevated intracellular sodium and calcium content, as well as an elevation of potassium-efflux under hyperthermia. Besides, hyperthermia has been demonstrated to induce various changes of cytoskeletal organization (cell shape, mitotic apparatus, endoplasmatic reticulum and lysosomes). The appearance of membrane blebbing in cultured cells was first described and then it was noted that cells underwent cell death after a single heat dose [2]. From a more recent point of view, membrane blebbing does not represent a primary damage of the cell membrane, but is a typical feature of apoptosis. The term apoptosis or programmed cell death defines a genetically encoded cell death program, which is morphological, biochemical and molecular distinct from necrosis [3]. This will be later described in more detail.

2.3. Cellular Proteins

Intracellular de novo synthesis and polymerization of both RNA- and DNA-molecules as well as protein synthesis are decreased in vitro at temperatures between 42 and 45°C in a dose-dependent manner. Whereas RNA- and protein synthesis recover rapidly after termination of heat exposure, DNA-synthesis is inhibited for a longer period [4-5]. Heat shock induces an aggregation of denatured proteins at the nuclear matrix. This is mainly due to insolubility of cellular proteins after heat-induced protein unfolding, entailing an enhancement of the nuclear protein concentration. Increase of the nuclear protein content by heat consequently affects several molecular functions (including DNA-synthesis and –repair) when a certain thermal dose is exceeded. This threshold dose is different between distinct cell lines. In the 1960s, hyperthermia was supposed to act similar as radiation by causing direct damage of DNA. Later, it became evident that heat is not able to cause severe DNA-damage by itself, but instead hinders the repair of induced cell damage, and thus boosts radiation-induced DNA-fragmentation. This may be caused by a temperature-dependant inhibition of DNA-repair enzymes.

Whereas synthesis of most cellular proteins is impaired under hyperthermic conditions there is one group of proteins, the so-called heat shock proteins (hsp), that are increasingly synthesized after heat application. Hsp represent a heterogeneous group of molecular chaperones consisting of at least five subgroups with different molecular mass and partially varying biological function [6]. Some are constitutively expressed and associated with specific intracellular organelles, and others are rapidly induced in response to cellular stress. The heat shock proteins are encoded by genes whose expression is substantially increased during stress conditions, such as heat shock, alcohol, inhibitors of energy metabolism, fever, inflammation etc. During these conditions, hsp increase cell survival by protecting and disaggregating stress-labile proteins, as well as proteolysis of the damaged proteins [7]. Under non-stress conditions, hsp have multiple housekeeping functions such as folding newly synthesized proteins, activation of regulatory proteins, protein signaling, etc [8]. Hsp-synthesis can be induced within minutes by activation of the so-called "heat shock factors". Cells recovering from a transient exposure to mild hyperthermia contain elevated levels of hsp and consequently possess an increased ability to tolerate normally lethal exposures to elevated temperatures, a phenomenon known as induced thermotolerance [9]. Furthermore, hsp are correlated with tumor genesis. For example, hsp are known to be over-expressed in several tumor cells [10-11] and to enhance the tumorgenicity of cells [12].

Another interesting feature of hsp comes from the interaction of the stress response with apoptosis. Interactions between these two pathways determine the fate of a cell and, as such, have a profound effect on the biological consequences of stress. Furthermore, data from various literature suggest a close connection between hsp-expression and inhibition of hyperthermic cell death, especially apoptosis, but the mechanisms involved are still poorly understood [13-14]. These preceding examples show how hsp are able to regulate the apoptotic process and are involved in tumor genesis indicating the therapeutic potential. Intervention in the synthesis or activity of hsp to modulate apoptosis therefore is an exciting novel therapeutic target in anticancer protocols.

2.4. Apoptosis

Apoptosis is a highly regulated cell death process, also called programmed cell death. Apoptotic cell death plays an important role in many different biological processes, like embryonic development, regulation of cell numbers and tissue homeostasis. Cells harbour both pro- and anti-apoptotic proteins and the balance between these proteins determines whether a cell is susceptible to undergo apoptosis. Disturbance of this balance can lead to pathological processes and disease. A reduced ability to undergo apoptosis may lead to the accumulation of cells that are potentially dangerous harbouring mitogenic aberrations, genomic instability or other alterations that contribute to cancer development. Many types of anticancer techniques cause damage to cancer cells that result in apoptotic cell death. Stimulation of apoptosis in malignant cells can therefore promote the susceptibility to anticancer treatments such as hyperthermia.

During apoptosis cells are orderly and carefully dismantled, without damage to the surrounding tissue. Morphological changes include condensation of nuclear chromatin, shrinkage of cells, increased cellular density and surface membrane blebbing [15]. The DNA is degraded in fragments of distinctive length by endonuclease activity. Membrane alterations (e.g. loss of asymmetric distribution of phosphatidylserines in the plasma membrane) occur that mark the dying cells for phagocytosis. Cellular remains are encapsulated in membranous vesicles, or apoptotic bodies. These apoptotic bodies are engulfed by phagocytes in the surrounding tissue, without causing an inflammatory reaction (figure 1).

Apoptosis is a stepwise process that is regulated by many protein families and cellular factors at different levels. The apoptotic process is initiated when a cell death signal is delivered to the cell. Subsequently the signal is transduced in biochemical pathways that eventually activate the execution of apoptotic cell disassembly through caspases. Two main biochemical pathways are known. The extrinsic or receptor-mediated pathway involves members of the TNF receptor (TNFR) superfamily and is engaged in response to cytokines and other extracellular signals. The intrinsic or mitochondria-mediated pathway is activated in response to intracellular signals and is controlled by members of the Bcl-2 family [16-17]. The mitochondria play a pivotal role in the intrinsic pathway of apoptosis [18]. The translocation of apoptotic proteins residing in the inter-membrane space requires permeabilization of the outer membrane. The mechanism by which this occurs is still subject of much debate.

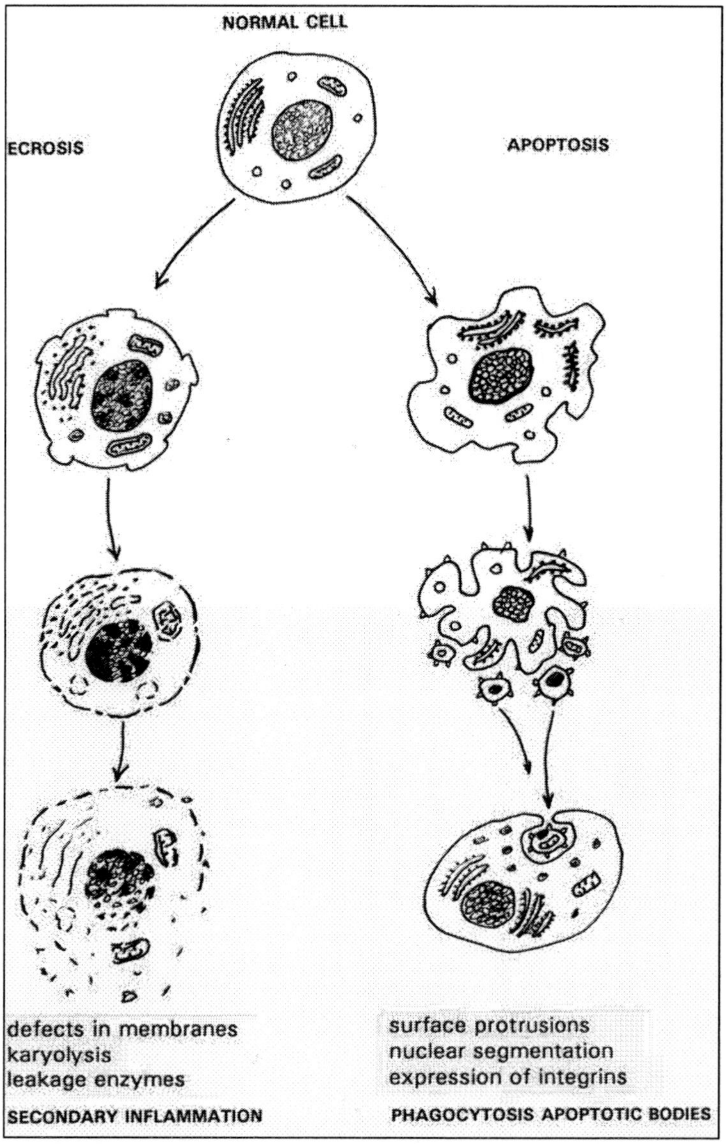

Figure 1. Representation of the morphological changes of a cell during apoptosis and necrosis. When the apoptotic death process is activated the dying cell undergoes several changes that alter its morphology. The cell shrinks and cellular density increases. The chromatin condensates and moves to the margins of the nucleus. DNA fragmentations occurs and the nucleus falls apart into fragments. The plasma membrane starts to bleb and sheds off vacuoles containing cellular remains called apoptotic bodies. When necrosis is induced the cell swells up without structural alterations which finally results in the rupture of the plasma membrane releasing the cellular content into the environment which can cause inflammatory reactions. Reproduced from [15].

Upon activation of the apoptotic process by death signals a highly regulated process of controlled events is followed. This process can be divided into 3 phases: 1) the initiation phase which involves the activation of surface death receptors or in the case of heat shock by the mitochondrial pathway; (2) the signal transduction phase, where activation of initiator caspases and certain kinases/phosphatases takes place, followed by (3) the execution phase which involves the activation of effector caspases. In response to stress signals, levels of the tumor suppressor protein p53 are rapidly increased [19], and its activity is enhanced after phosphorylation at the Ser-15 residu, resulting in the up-regulation of downstream genes, including the proapoptotic Bax protein. In turn, increased levels of Bax induce mitochondrial membrane permeabilization (MOMP), release of cytochrome c and activation of a caspase cascade, leading to apoptosis (see figure 2).

Hyperthermia is suitable to induce both necrotic and programmed cell-death in a temperature-dependent manner [20]. Programmed cell death seems to represent an important effector of heat action. The susceptibility of cultured cells to apoptosis due to hyperthermia could be demonstrated in various studies. Thereby, the response varied from different cell-lines, temperatures, duration, etc [21]. That many different types of cells die by apoptosis in response to mild hyperthermia, is a well established fact; what remains in question are the mechanisms that trigger and affect the apoptotic response, leading to cell death. With regard to hyperthermia heat-shock response represents a typical, but by no means specific reaction. Hsp have an extremely complex role in the regulation of apoptosis. Historically, studies on the protective ability of the hsp have been focused largely on their role as chaperones to prevent misfolding of proteins and to accelerate their refolding. However, more recently, the function of hsp has been shown to be broader and encompass an anti-apoptotic role that can, but does not always, depend upon their chaperone activity [22]. Moreover, hsp are found to be of importance in fundamental processes in apoptosis. Hsp are able to inhibit heat-induced apoptosis upstream of mitochondria by preventing Bax translocation [23] and downstream mitochondria by inhibiting the release of pro-apoptotic factors from mitochondria following stress [24]. Recently, it was shown that the multidomain proapoptotic molecules Bax and Bak are directly activated by heat [25] and that caspase-2 plays an important role in heat-induced apoptosis via cleavage of Bid [26].

Hsp70 is also known to interact with the p53 protein. As a response to DNA damage by exposure to many stress signals, including heat shock, tumor suppressor protein p53 is activated and cell cycle arrest is promoted. However, when the stress is excessive, p53 can induce tumor-suppressive apoptotic cell death. Mutation of the tumor suppressor protein p53 gene is one of the major mechanisms of tumor cells to escape from apoptosis. The function of p53 is recently correlated with hsp [27]. This interaction results in stabilization of p53-DNA binding conformation and localization of p53 from the cytoplasm to the nucleus. Recent studies to the interaction of hsp and p53 revealed that expression of functional wt p53 together with hsp are able to protect cells from heat-induced apoptosis [28], whereas over expressed hsp with mutated p53 did not show this protection [29].

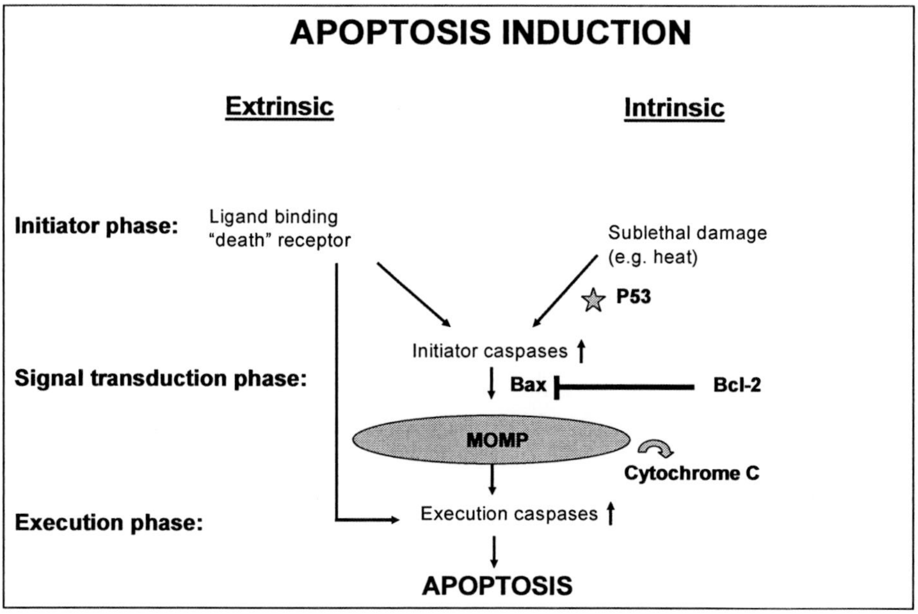

Figure 2. Simplified scheme of the apoptosis induction mechanism. Cell death stimuli can be delivered to a cell as extracellular or intracellular signals. In the extrinsic pathway, extracellular ligands can bind to "death" receptors at the cell membrane. Intrinsic or extrinsic apoptotic stimuli leads to the permeabilization of the mitochondrial membrane regulated by the pro-apoptotic Bax and the anti-apoptotic Bcl-2. this results in the leakage of apoptotic substrates like cytochrome c and the activation of effector caspases leading to apoptosis.

3. Applying Hyperthermia

Several difficulties or obstacles arise when trying to localize and control heating to tumor sites. The actual dimensions or geometry of tumors are often irregularly shaped in both lateral extent and depth dimensions. Furthermore, heterogeneous distribution of tumor blood vessels (especially in larger tumors) results in regions with low blood perfusion and highly vascularized, well-perfused zones. The resulting heating patterns are often "cold" in the well-perfused zones with "hot" spots in the regions with low blood flow. With many of the current heating technologies, the required temperatures and thermal doses throughout the complete tumor volume have been difficult to achieve in anything but small superficial tumors. Mediocre clinical results in the past have been attributed to the use of inadequate heating techniques that have produced temperatures which were not well controlled and inhomogeneously distributed.

Until recent years, electromagnetic (EM) heating techniques have been the most frequently used approach for applying conventional hyperthermia treatments. There are certain positive attributes associated with the EM approach, but the large wavelengths and/or high energy absorption rates in tissue make it difficult to control and localize energy deposition. Thus, this approach has been limited mostly to trying very superficial disease or regional heating of deep sites. Due to the increasing overlap of different fields of science some interesting new approaches have arised. Ultrasound technology has inherent properties that can potentially be used to overcome or compensate for the difficulties mentioned above and provide increased localization and control of temperature distributions as required for thermal therapies. The key-words in these novel combinational techniques are specificity and selectivity. When heat can be applied in a controlled way to sites within the body or even cell type specific, cancer therapy can be selective for tumor cells with minimal damage to surrounding healthy tissue in a non-invasive manner. New interesting techniques include therapeutic possibilities for selective cell targeting with infrared-absorbing gold-nanoparticles. Another approach to increase the effectiveness of hyperthermia as conventional cancer treatment modality is to interfere with molecular mechanisms regarding controlled cell death (apoptosis) either by stimulating pro-apoptotic or blocking anti-apoptotic players in the apoptotic cell death mechanism. Hereby, selective integration of drugs or genes into tumor cells is preferable. The use of microbubbles in combination with ultrasound has properties that can potentially be used to combine hyperthermia with the uptake of drugs or genes resulting in increased apoptosis. In this chapter, the use of ultrasound and EM heating techniques as well as implementation of new approaches are discussed.

3.1. Electromagnetic Field

There is a considerable interest in cellular responses to electromagnetic fields (EMF) largely due to the daily use of equipment that generates EMF in the radiowave and microwave regions. Over the recent decades, it has been stated that the most important effect of EMF on biological systems is heat development [31]. EM energy in these frequencies can readily be transmitted through, absorbed by and reflected from the tissue of the body. The temperature rise in the tissue is due to the transfer of EM energy into heat. Frictional forces between water molecules in the tissue are generated as the EM waves pass through the tissue and the water molecules are vibrated. This friction results in heat production and by concentrating or focussing these EM waves significant increase in temperature can be obtained. Because EMF cause water molecules to vibrate, differences in water content within tissues (fat, bone, muscle etc.) determine the extent of temperature increase by EMF applicators. In order to produce the required heating in a specific region the dielectric and conductive properties of the biological material must be known [32]. In this way the energy can be adjusted to achieve efficient energy absorption and thereby predict the amount of heating. The temperature dependence of both conductivity and dielectric constant is related mainly to the water content [33]. EMF arise from the motion of electric charges. The strength of the magnetic field is measured in amperes per meter (A/m); more commonly in research the quantity related flux density (in microtesla, μT) is being used. The most commonly used frequencies for hyperthermic applications are just above 300 MHz (microwave range) [34]. The effects of EMF on cell lines have been extensively studied, however besides the thermal effect contradictory reports have been published about the non-thermal effects.

Extremely low frequencies (ELF) in the order of 50Hz were found to induce apoptosis in vitro in several studies [35] however, studies that failed to correlate ELF with induced apoptosis [36] have also been published. For the use of high-frequency EMF in the order of

GHz it has also been stated that interactions with cellular components are possible, nevertheless it has never been proven. This discrepancy of results in research towards effects of EMF on biological material is mainly due to the enormous complexity of exposure-response relationships for non-thermal effects. Furthermore, inconsistent results published in the literature may be caused by the variability of exposure systems, the exposure conditions and the cell types used. This makes it very difficult to compare the results of different studies describing the biological effects of EMF on cells.

Although the effects of electromagnetic energy are still poorly understood, hyperthermia induced by EMF is a main method for such non-invasive heating. The advantages of EMF heating are that EM energy can propagate through air. Thus, due to the presence of air within and in the vicinity of lungs, stomach, bowel, bladder etc, the use of the EM technique is suggested for therapy in these regions. EM energy can penetrate bones and therefore be used for applications in the chest area. These advantages of EM radiation are mainly due to strong absorption in water-containing tissues. Therefore, there are potential hazards for the EM technique in tissues close to organs containing or surrounded by fluids such as the heart, stomach etc.

3.1.1. Nanoparticles

To overcome problems of overheating tumor-surrounding tissue, between the source and the target site, and to sufficiently penetrate the tumor site with the appropriate heating temperature new techniques were investigated. Experimental investigations of the application of magnetic materials for hyperthermia dates back to 1957 when Gilchrist et al [37] heated various tissue samples with 20-100nm size particles of γ-Fe_2O_3 exposed to a 1.2 MHz electromagnetic field. Since then there have been numerous publications describing a variety of schemes using different types of magnetic materials, different field strengths and frequencies and different methods of encapsulation and delivery of the particles [38-40]. Generally, the procedure involves dispersing magnetic particles throughout the target tissue, and then applying an electromagnetic field of sufficient strength and frequency to cause the particles to heat. A number of studies have demonstrated the therapeutic efficacy of this form of treatment in animal models [41].

To date, however, there have been no reports of the successful application of this technology to the treatment of a human patient. The challenge lies in being able to deliver an adequate quantity of the magnetic particles to generate enough heat in the target using magnetic fields that are clinically acceptable. Furthermore, the frequency and the strength of the externally applied electromagnetic field are limited by physiological responses to high frequency magnetic fields such as muscle spasm, possible cardiac stimulation and arrhythmia [41]. There are different methods of administration; direct injection, intravascular or antibody targeting.

Since the first ferromagnetic microspheres of the 1970s, a variety of magnetic nano- and microparticles have been developed [42-43]. The optimization continues today. Generally, the magnetic component of the particle is coated by a biocompatible polymer such as PVA or dextran, although recently inorganic coatings such as silicia have been developed. The coatings act to shield the magnetic particle from the surrounding environment and can also be modified by functional groups and antibodies against tumor specific cells [44]. The use of antibodies specific for a tumor cell type enables precise and specific targeting the tumor region thereby minimizing the effect on the surrounding tissue [45-46]. Regarding the choice of magnetic particles, the iron oxide magnetite (Fe_3O_4) and maghemite (γ-Fe_2O_3) are the most studied to date because of their generally appropriate magnetic properties and biological compatibility, although many others have been investigated. Particle sizes less than about 10 μm are normally considered small enough to enable effective delivery to the site of the cancer, either via encapsulation in a larger moiety or suspension in some kind of carrier fluid. Research also continues into alternative magnetic particles, such as iron, cobalt or nickel. Since magnetic particles have unique magnetic features, they can be applied to special medical techniques such as immunoassays, magnetic resonance imaging (MRI), drug/gene delivery and hyperthermia [47-50].furthermore, the combination of hyperthermia by nanoparticles together with MRI and/or drug delivery techniques may offer attractive possibilities in biomedicine.

Recently new in this field are the gold nanoparticles, which are in the size of 10-100 nm and are able to undergo plamon resonance with light. The first account of the use of gold particles in hyperthermal therapy was published in 2003. Thereby, Halas et al used gold-on-silica nanoshells to target breast carcinoma cells using the HER2 antibody. By adjusting the relative core and shell thickness, nanoshells can be manufactured to absorb or scatter light at a desired wave length across visible and near infrared (NIR) wavelengths. This optical tunability permits the fabrication of nanoshells with a peak optimal absorption in the NIR, a region of light where optical penetration through tissue is optimal. Because the body is moderately transparent to NIR light this provides an opportunity for therapeutic effects in deep tissues. O'Neal et al 2004 [51] reported impressive results in vivo by showing selective photo-thermal ablation in mice using near infrared-absorbing nanoparticles (see figure 3).

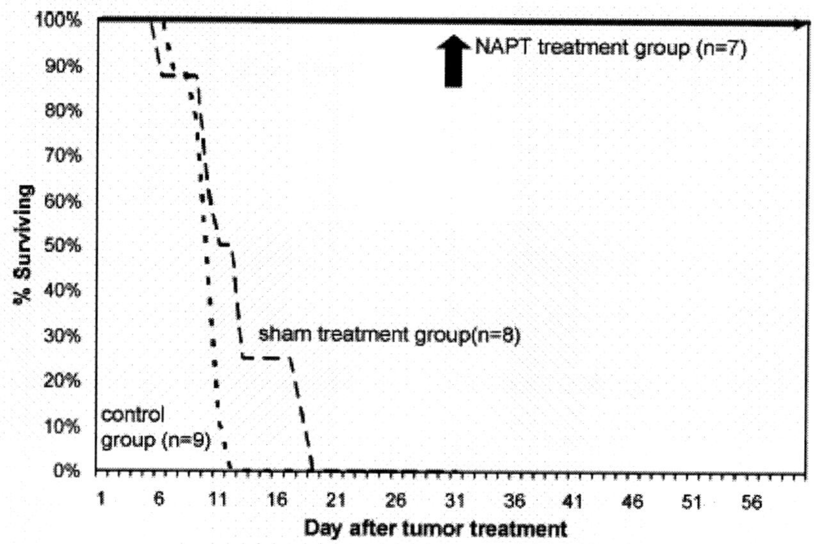

Figure 3. Photo-thermal tumor ablation in mice using near infrared-absorbing nanoparticles. A survival time plot for three experimental groups for 60 days. For the nanoshell-assisted photo-thermal therapy (NAPT) group the survival fraction was 100% after 60 days. For the sham-treated and control group the survival times were significantly lower with 10.1 and 12.5 days, respectively. Reproduced from [51].

3.2. Ultrasound

The interaction between ultrasound waves and living tissues was described en studied 70 years ago, and was first suggested as a form of therapy. Ultrasound has since been used to treat a variety of disorders, from skin wounds to malignant tumors.

Ultrasound consists of inaudible high-frequency mechanical vibrations created when a generator produces electrical energy that is converted to acoustic energy within the transducer. The waves produced are transmitted by propagation through molecular collision and vibration, with a progressive loss of the intensity of energy during passage through the tissue (attenuation), due to absorption, dispersion or scattering of the waves [52-53]. The total amount of energy in an ultrasound beam is its power expressed in watts. The amount of energy that reaches a specific site is dependent upon characteristics of the ultrasound (frequency, intensity, amplitude, focus, and beam uniformity) and the tissue through which it travels.

Therapeutic ultrasound has a frequency range of 0.75-3 MHz, with most machines set at a frequency of 1-3 MHz. Low-frequency ultrasound waves have a greater depth of penetration but are less focused. Ultrasound at a frequency of 1 MHz is absorbed primarily by tissue at depth of 3-5 cm and is therefore recommended for deeper injuries and in patients with more subcutaneous fat. A frequency of 3 MHz is recommended for more superficial lesions at depths of 1-2 cm. Tissues can be characterized by their acoustic impedance, the product of their density and the speed at which ultrasound travels through it. Low absorption (and therefore high penetration) of ultrasound is seen in tissue that are high in water content (e.g. fat), whereas absorption is higher in tissues rich in protein (e.g. skeletal muscle). The larger the difference in acoustic impedance between the different tissues, the less the transmission from one to another. When reflected ultrasound meets further waves being transmitted, a standing wave or hot spot may be created, which has a potential adverse effect on tissue. Coupling media between the interface of the transducer-head and the tissue prevent reflection of the waves away from the treated area by excluding air from the transducer and the tissue. Therapeutic ultrasound can be pulsed or continuous. The former has on/off cycles, each component of which can be varied to alter the dose. Continuous ultrasound has a greater heating effect than exposure to pulsed ultrasound.

The interaction between ultrasound and biological matters has been investigated by a number of authors for almost half a century [54-55]. However, the widespread use of therapeutic ultrasound in clinical environments has so far been limited, in part, due to incomplete understanding of the interaction process in vivo. Cellular effects have been studied in isolated cells, cell suspensions and-or cell cultures as well in cells in their natural setting in vivo. Studies of cells in culture provide a means for examining the effect of ultrasound without numerous biological variations operating in the whole organism. However, such simplification reduces the applicability of the experimental data to the actual situation in vivo [55]. Ultrasound of low intensities has no ionizing ability and its direct action on biological systems is effected by means of two mechanisms divided in thermal en non-thermal effects [56-57]. However, the idea rises that the reality is that the two effects are not separable. Generally, bioeffects of ultrasound exposure are intensity and frequency dependent. A higher intensity benefits heat production, and a lower frequency favours the occurrence of non-thermal effects. Therefore, the acoustic parameters must be selected carefully according to the objective required.

3.2.1. Thermal and Non-Thermal Effects

Although there is evidence for ultrasound exposure causing a rise in temperature, the extent of tissue heating is dependent on a number of variables. Heating is intensity dependent. The most used frequency in literature found for inducing hyperthermia is 1 MHz. Continues ultrasound at intensities ranging from 0.5-3 W/cm^2 usually is generally used to increase the temperature in the region/tissue of interest to hyperthermia temperatures (42-45°C). The higher the intensity the more heat will be formed and the more non-thermal functional alterations occur [58]. The time to heat the tissue depends on a lot of variables including intensity, frequency tissue characteristics etc. In vitro research studying the effects of ultrasound exposure therefore is highly dependent on the experimental set-up. As a consequence comparing data from different studies is a difficult task. To a large extent the effects of ultrasound on biological systems have been examined in in vitro studies. There is, however, little evidence that these changes occur in vivo. Extrapolating results from in vitro to in vivo conditions therefore is conjectural. In order to assess the effect of ultrasound on an intact organism, the influence of regulatory mechanisms such as homeostasis must be taken into account.

Therapeutic ultrasound produces a combination of non-thermals effects (acoustic streaming and cavitation) that are difficult to isolate. Acoustic streaming is defined as the physical forces of the sound waves that provide a driving force capable of displacing ions and small molecules. This also occurs at the cellular level with small organelles and molecules inflicting structural changes. Microstreaming is the only type of acoustic streaming with sufficient strength to alter membrane permeability and stimulate cell activity when it occurs at the boundary of the cell membrane and tissue fluid [55]. Cavitation is defined as the physical forces of the sound waves on micro- environmental gases and fluid. As the sound waves propagate through the medium, the characteristic compression and rarefaction cause microscopic gas bubbles in the tissue fluid to expand, contract and expand. It is generally thought that the rapid changes in pressure , both in and around the cell may cause damage to the cell. The exposure of biological tissues to ultrasound can result in structural and/or functional alterations. Structural changes range from slight but repairable damage to immediate death. The functional alterations include proliferation, migration, synthesis, secretion, gene expression and membranous action, etc [59-62].

3.2.2. Microbubbles

While microbubbles have already been used as contrast agents for a couple of decades, the use of microbubbles for therapy is in its infancy. In diagnostic ultrasound, microbubbles create an acoustic impedance mismatch between fluids and tissues to increase reflection of sound. The bubble wall typically consists of a 2-3 nm thick lipid monolayer and the inside can be filled with gas or air. However, air dissolves very rapidly in aqueous environments. Therefore perfluorochemicals when used as the filling gas or part of it, delay bubble dissolution very effectively, due to very low water solubility [63]. Microbubbles have extensive therapeutic applications beyond diagnosis. Not only do microbubbles increase reflection of sound, they also increase the absorption of sonic energy. As ultrasound interacts with the microbubbles, the microbubbles begin to oscillate or resonate. Eventually, depending upon acoustic power and other factors, the microbubbles will be destroyed by the ultrasound energy, and in doing so create a local shock wave [64]. Therefore microbubbles can be used as a controllable delivery system for drug or gene delivery [65]. This technique in combination with so-called 'smart' microbubbles, whereby specific tumor receptors can be targeted, significantly increases the specificity of the delivery system. Active targeting includes attachment of receptor ligands, including monoclonal antibodies, polysaccharides and peptides that recognize disease antigens to the microbubble surface. Destruction of the microbubbles by the use of high intensity ultrasound increases capillary permeability and delivery of material to the interstitial tissue. A variety of different drugs and genes may potentially be incorporated into the microbubbles. For example genes that increase apoptosis or inhibit proliferation of tumor cells.

The combination of using ultrasound for hyperthermia induction combined with the different therapeutic strategies can be a powerful tool to eradicate tumor cells. Several strategies have been used to direct microbubbles to the region of disease. In this way the microbubbles bind specifically to the tumor antigen of interest and by activating the bubbles their content is delivered into the cell. Selective retention of microbubbles in regions of disease has been achieved by conjugating ligands to the surface of ultrasound contrast agents. For the imaging of e.g. inflammation, monoclonal antibodies that recognize endothelial cell-adhesion molecules have been conjugated to the surface of microbubbles or air-containing liposomes [66-68]. These microbubbles have been used for example to image vascular phenotype in ischaemia-reperfusion injury and atherosclerotic disease. In the same way microbubbles might be used to target specific tumor cells.

4. Conclusion

Although the therapeutic relevance of hyperthermia is already known for decades, the technique is still far from being used as a routine anti-cancer therapy, despite the advantages this technique offers by means of minimal side-effects and its non-invasive character. This is mainly because of inadequate heating of the tumors and poor understanding of the molecular processes involved in heat-induced apoptosis. The ways to induce hyperthermia are numerous, all with their own advantage and disadvantages.

The effects of EMF on biosystems are well studied in the past decades, mainly because of the increasing contact to EMF's in modern society due to mobile phones, computers etc. The outcomes of these studies, however, are not consistent. Although the thermal effects of EMF on living tissue are well documented and therefore are not an issue, the non-thermal effects are. This is mainly due to the enormous complexity of exposure-response relationships for non-thermal effects. Furthermore, inconsistent results published in the literature may be caused by the variability of exposure systems, the exposure conditions and the cell types used emphasizing the need for a controllable and validated system. A new and potential promising application for the use of EMF in the hyperthermia field is the combination of EMF and nanoparticles. The use of nanoparticles enables a more specific heating of the tumor site resulting in fewer side effects. Another possibility of using nanoparticles is the combination with other medical techniques such as MRI and drug delivery.

The interaction between ultrasound and biological matters has been investigated by a number of authors for almost half a century. However, the widespread use of therapeutic ultrasound in clinical environments has so far been limited, in part, due to incomplete understanding of the interaction process in vivo. In vitro ultrasound is known to cause thermal as well as non-thermal effects. These non-thermal effects caused by cavitation and other mechanical effects generally are believed to be responsible for the cell killing effect of ultrasound seen in vitro. There is however little evidence that these changes also occur in vivo, extrapolating data from in vitro to in vivo environment therefore is highly speculative. Gas-filled microbubbles are being used as contrast agents for ultrasound already. A relatively new field is the use of these microbubbles in a therapeutic way. Ultrasound causes the microbubbles to be destroyed by acoustic impedance, creating a shock wave. These characteristics of the microbubbles enables drug/ and or gene delivery at the site of interest by the use of ultrasound. In combination with hyperthermia this can be a powerful tool.

Recently, direct targeting of tumor cells, thereby sparing healthy tissue, has been investigated resulting in new exciting techniques with great potential such as selective cell-targeting with nanoparticles. This, together with new insights into the molecular mechanisms involved in heat-induced apoptosis and thermotolerance, provides tools to overcome problems faced in the past. As more aspects of heat-induced apoptosis are being elucidated, the effect of heat on cells has proven to be very complex. Interactions of the ancient defense mechanisms of hsp with the apoptotic cascade are now a generally accepted dogma and new therapeutic strategies arise from this point of view. Hsp can be used as novel molecular targets for pharmacological and therapeutic interventions both to prevent and to initiate apoptosis. These strategies entail the increase of the heat-induced effect on cells by breaking down the defense mechanism of the cell by inhibiting or interfering with hsp in order to increase hyperthermia-induced apoptosis. For future research to get a full picture of hyperthermia-induced apoptosis more studies must be performed not only in vitro but together with novel heating strategies e.g. using nanoparticles in vivo to implement the fundamental knowledge to therapeutic settings.

References

1. Narita K, Hanakawa K, Kasahara T, Hisamitsu T, Asano K. (1997). Induction of apoptotic cell death in human leukemic cell line, HL-60, by extremely low frequency electric magnetic fields: analysis of the possible mechanisms in vitro. *In Vivo, 11*, 329-35.
2. Borrelli MJ, Wong RS, Dewey WC. (1986). A direct correlation between hyperthermia-induced membrane blebbing and survival in synchronous G1 CHO cells. *J Cell Physiol, 126(2)*, 181-90.
3. Vermes I, Haanen C, Reutelingsperger C. (2000). Flow cytometry of apoptotic cell death. *J Immunol Methods, 243(1-2)*, 167-90.

4. Streffer C. (1988). Aspects of metabolic change after hyperthermia. *Recent Results Cancer Res, 107*, 7-16.
5. Falk MH, Issels RD. (2001). Hyperthermia in oncology. *Int J Hyperthermia, 17(1)*, 1-18.
6. Sreedhar AS, Csermely P. (2004). Heat shock proteins in the regulation of apoptosis: new strategies in tumor therapy: a comprehensive review. *Pharmacol Ther, 101(3)*, 227-57.
7. Parsell DA, Lindquist S. (1993). The function of heat-shock proteins in stress tolerance: degradation and reactivation of damaged proteins. *Annu Rev Genet, 27*, 437-96.
8. Samali A, Holmberg CI, Sistonen L, Orrenius S. (1999). Thermotolerance and cell death are distinct cellular responses to stress: dependence on heat shock proteins. *FEBS Lett, 461(3)*, 306-10.
9. Urano M. (1989). Kinetics of thermotolerance in normal and tumor tissues: a review.
Cancer Res, 46(2), 474-82.
10. Soti C, Csermely P. (1998). Molecular chaperones in the etiology and therapy of cancer. *Pathol Oncol Res, 4(4)*, 316-21.
11. Jolly C, Morimoto RI. (1999). Stress and the cell nucleus: dynamics of gene expression and structural reorganization. *Gene Expr, 7(4-6)*, 261-70.
12. Garrido C, Fromentin A, Bonnotte B, Favre N, Moutet M, Arrigo AP, Mehlen P, Solary E. Heat shock protein 27 enhances the tumorigenicity of immunogenic rat colon carcinoma cell clones. *Cancer Res. 1998, 58(23)*, 5495-9.
13. Buchner J. (1996). Supervising the fold: functional principles of molecular chaperones. *FASEB J, 10(1)*, 10-9.
14. Kaur J, Kaur J, Ralhan R. (2000). Induction of apoptosis by abrogation of HSP70 expression in human oral cancer cells. *Int J Cancer, 85(1)*, 1-5.
15. Vermes I, Haanen C. (1994). Apoptosis and programmed cell death in health and disease.
Adv Clin Chem, 31, 177-246.
16. Kim R, Emi M, Tanabe K. (2006). Role of mitochondria as the gardens of cell death.
Cancer Chemother Pharmacol, 57(5), 545-53.
17. Dejean LM, Martinez-Caballero S, Manon S, Kinnally KW. Regulation of the mitochondrial apoptosis-induced channel, MAC, by BCL-2 family proteins. *Biochim Biophys Acta.* 2006, 1762(2), 191-201.
18. Armstrong JS. (2006). Mitochondrial membrane permeabilization: the sine qua non for cell death. *Bioessays, 28(3)*, 253-60.
19. Striteska D. (2005). The tumor supressor gene p53. *Acta Medica, 48(1)*, 21-5.
20. Harmon BV, Corder AM, Collins RJ, Gobe GC, Allen J, Allan DJ, Kerr JF. (1990). Cell death induced in a murine mastocytoma by 42-47 degrees C heating in vitro: evidence that the form of death changes from apoptosis to necrosis above a critical heat load. *Int J Radiat Biol, 58(5)*, 845-58.

21. O'Neill KL, Fairbairn DW, Smith MJ, Poe BS. (1998). Critical parameters influencing hyperthermia-induced apoptosis in human lymphoid cell lines. *Apoptosis, 3(5)*, 369-75.
22. Beere HM. (2004). "The stress of dying": the role of heat shock proteins in the regulation of apoptosis. *J Cell Sci, 117*, 2641-51.
23. Stankiewicz AR, Lachapelle G, Foo CP, Radicioni SM, Mosser DD. (2005). Hsp70 inhibits heat-induced apoptosis upstream of mitochondria by preventing Bax translocation.
J Biol Chem, 280(46), 38729-39.
24. Mosser DD, Caron AW, Bourget L, Meriin AB, Sherman MY, Morimoto RI, Massie B. The chaperone function of hsp70 is required for protection against stress-induced apoptosis. *Mol Cell Biol, 20(19)*, 7146-59.
25. Pagliari LJ, Kuwana T, Bonzon C, Newmeyer DD, Tu S, Beere HM, Green DR. (2005). The multidomain proapoptotic molecules Bax and Bak are directly activated by heat. Proc *Natl Acad Sci U S A, 102(50)*, 17975-80.
26. Bonzon C, Bouchier-Hayes L, Pagliari LJ, Green DR, Newmeyer DD. (2006). Caspase-2-induced apoptosis requires bid cleavage: a physiological role for bid in heat shock-induced death. *Mol Biol Cell, 17(5)*, 2150-7.
27. Zylicz M, King FW, Wawrzynow A. (2001). Hsp70 interactions with the p53 tumour suppressor protein. *EMBO J, 20(17)*, 4634-8.
28. Sreedhar AS, Pardhasaradhi BV, Khar A, Srinivas UK. (2002). A cross talk between cellular signalling and cellular redox state during heat-induced apoptosis in a rat histiocytoma. *Free Radic Biol Med, 32(3)*, 221-7.
29. Sreedhar AS, Pardhasaradhi BV, Khar A, Srinivas UK. (2002). Effect of C-terminal deletion of P53 on heat induced CD95 expression and apoptosis in a rat histiocytoma.
Oncogene, (25), 4042-9.
30. Anhalt DP, Hynynen K, Roemer RB. (2003). Patterns of changes of tumour temperatures during clinical hyperthermia: implications for treatment planning, evaluation and control.
Int J Hyperthermia, 11(3), 425-36.
31. Stacey M, Stickley J, Fox P, Statler V, Schoenbach K, Beebe SJ, Buescher S. (2003). Differential effects in cells exposed to ultra-short, high intensity electric fields: cell survival, DNA damage, and cell cycle analysis. *Mutat Res, 542(1-2)*, 65-75.
32. Vrba J, Lapes M, Oppl L. (1999). Technical aspects of microwave thermotherapy. *Bioelectrochem Bioenerg, 48(2)*, 305-9.
33. Watmough and W.M. Ross (Blackie and Son, Glasgow, 1986) *Hyperthermia*.
34. Rietveld PJ, van Putten WL, van der Zee J, van Rhoon GC. (1999). Comparison of the clinical effectiveness of the 433 MHz Lucite cone applicator with that of a conventional waveguide applicator in applications of superficial hyperthermia. *Int J Radiat Oncol Biol Phys, 43(3)*, 681-7.
35. Narita K, Hanakawa K, Kasahara T, Hisamitsu T, Asano K. (1997). Induction of apoptotic cell death in human leukemic cell line, HL-60, by extremely low

frequency electric magnetic fields: analysis of the possible mechanisms in vitro. *In Vivo, 11(4)*, 329-35.

36. Ding GR, Nakahara T, Hirose H, Koyama S, Takashima Y, Miyakoshi J. (2004). Extremely low frequency magnetic fields and the promotion of H2O2-induced cell death in HL-60 cells. *Int J Radiat Biol, 80(4)*, 317-24.
37. Gilchrist R.K., Meadl R., Shorey W.D., Hanselman RC, Parrot JC, Taylor CB. (1957). Selective inductive heating of lymph nodes. *Ann Surg, 146(4)*, 596-606.
38. Minamimura T, Sato H, Kasaoka S, Saito T, Ishizawa S, Takemori S, Tazawa K, Tsukada K. (2000). Tumor regression by inductive hyperthermia combined with hepatic embolization using dextran magnetite-incorporated microspheres in rats. *Int J Oncol, 16(6)*, 1153-8.
39. Moroz P, Jones SK, Winter J, Gray BN. (2001). Targeting liver tumors with hyperthermia: ferromagnetic embolization in a rabbit liver tumor model. *J Surg Oncol, 78(1)*, 22-9.
40. Jones SK, Winter JG, Gray BN. (2002). Treatment of experimental rabbit liver tumours by selectively targeted hyperthermia. *Int J Hyperthermia, 18(2)*, 117-28.
41. Moroz P, Jones SK, Gray BN. (2002). Magnetically mediated hyperthermia: current status and future directions. *Int J Hyperthermia, 18(4)*, 267-84.
42. Bettge M, Chatterjee J, Haik Y. (2004). Physically synthesized Ni-Cu nanoparticles for magnetic hyperthermia. *Biomagn Res Technol, 2(1)*, 4.
43. Ito A, Kuga Y, Honda H, Kikkawa H, Horiuchi A, Watanabe Y, Kobayashi T. (2004). Magnetite nanoparticle-loaded anti-HER2 immunoliposomes for combination of antibody therapy with hyperthermia. *Cancer Lett, 212(2)*, 167-75.
44. Shinkai M, Ito A. (2004). Functional magnetic particles for medical application. *Adv Biochem Eng Biotechnol, 91*, 191-220.
45. El-Sayed IH, Huang X, El-Sayed MA. (2006). Selective laser photo-thermal therapy of epithelial carcinoma using anti-EGFR antibody conjugated gold nanoparticles. *Cancer Lett, 239(1)*, 129-35.
46. Ito A, Tanaka K, Kondo K, Shinkai M, Honda H, Matsumoto K, Saida T, Kobayashi T. (2003). Tumor regression by combined immunotherapy and hyperthermia using magnetic nanoparticles in an experimental subcutaneous murine melanoma. *Cancer Sci, 94(3)*, 308-13.
47. Ito A, Matsuoka F, Honda H, Kobayashi T. (2003). Heat shock protein 70 gene therapy combined with hyperthermia using magnetic nanoparticles. *Cancer Gene Ther, 10(12)*, 918-25.
48. Ma N, Szmitko P, Brade A, Chu I, Lo A, Woodgett J, Klamut H, Liu FF.(2004). Kinase-dead PKB gene therapy combined with hyperthermia for human breast cancer. *Cancer Gene Ther, 11(1)*, 52-60.
49. Huang Q, Hu JK, Lohr F, Zhang L, Braun R, Lanzen J, Little JB, Dewhirst MW, Li CY. (2000). Heat-induced gene expression as a novel targeted cancer gene therapy strategy. *Cancer Res, 60(13)*, 3435-9.
50. Lebedeva IV, Su ZZ, Sarkar D, Fisher PB. (2003). Restoring apoptosis as a strategy for cancer gene therapy: focus on p53 and mda-7. Semin *Cancer Biol, 13(2)*, 169-78.

51. O'Neal DP, Hirsch LR, Halas NJ, Payne JD, West JL. (2004). Photo-thermal tumor ablation in mice using near infrared-absorbing nanoparticles. *Cancer Lett, 209(2)*, 171-6.
52. Colombel M, Gelet A. (2004). Principles and results of high-intensity focused ultrasound for localized prostate cancer. *Prostate Cancer Prostatic Dis, 7(4)*, 289-94.
53. Barnett SB, Rott HD, ter Haar GR, Ziskin MC, Maeda K. (1997). The sensitivity of biological tissue to ultrasound. *Ultrasound Med Biol, 23(6)*, 805-12.
54. Dalecki D. (2004). Mechanical bioeffects of ultrasound. *Annu Rev Biomed Eng, 6*, 229-48.
55. Hrazdira I, Skorpikova J, Dolnikova M. (1998). Ultrasonically induced alterations of cultured tumour cells. *Eur J Ultrasound, 8(1)*, 43-9.
56. ter Haar G. (1999). Therapeutic ultrasound. *Eur J Ultrasound, 9(1)*, 3-9.
57. Dalecki D. (2004). Mechanical bioeffects of ultrasound. *Annu Rev Biomed Eng, 6*, 229-48.
58. Baker KG, Robertson VJ, Duck FA. (2001). A review of therapeutic ultrasound: biophysical effects. *Phys Ther, 81(7)*, 1351-8.
59. Reher P, Doan N, Bradnock B, Meghji S, Harris M. (1999). Effect of ultrasound on the production of IL-8, basic FGF and VEGF. *Cytokine, 11(6)*, 416-23.
60. Liu Y, Cho CW, Yan X, Henthorn TK, Lillehei KO, Cobb WN, Ng KY. (2001). Ultrasound-Induced hyperthermia increases cellular uptake and cytotoxicity of P-glycoprotein substrates in multi-drug resistant cells. *Pharm Res, 18(9)*, 1255-61.
61. Cho CW, Liu Y, Cobb WN, Henthorn TK, Lillehei K, Christians U, Ng KY. (2002). Ultrasound-induced mild hyperthermia as a novel approach to increase drug uptake in brain microvessel endothelial cells. *Pharm Res, 19(8)*, 1123-9.
62. Honda H, Kondo T, Zhao QL, Feril LB Jr, Kitagawa H. (2004). Role of intracellular calcium ions and reactive oxygen species in apoptosis induced by ultrasound. *Ultrasound Med Biol, 30(5)*, 683-92.
63. Pitt WG, Husseini GA, Staples BJ. (2004). Ultrasonic drug delivery: a general review. *Expert Opin Drug Deliv, 1(1)*, 37-56.
64. Unger EC, Matsunaga TO, McCreery T, Schumann P, Sweitzer R, Quigley R. (2002). Therapeutic applications of microbubbles. *Eur J Radiol, 42(2)*, 160-8.
65. El-Aneed A. (2004). Current strategies in cancer gene therapy. *Eur J Pharmacol, 498(1-3)*, 1-8.
66. Weller GE, Lu E, Csikari MM, Klibanov AL, Fischer D, Wagner WR, Villanueva FS. (2003). Ultrasound imaging of acute cardiac transplant rejection with microbubbles targeted to intercellular adhesion molecule-1. *Circulation, 108(2)*, 218-24.
67. Demos SM, Alkan-Onyuksel H, Kane BJ, Ramani K, Nagaraj A, Greene R, Klegerman M, McPherson DD. (1999). In vivo targeting of acoustically reflective liposomes for intravascular and transvascular ultrasonic enhancement. *J Am Coll Cardiol, 33(3)*, 867-75.

68. Lindner JR, Song J, Christiansen J, Klibanov AL, Xu F, Ley K. (2001). Ultrasound assessment of inflammation and renal tissue injury with microbubbles targeted to P-selectin. *Circulation, 104(17)*, 2107-12.

In: Cell Apoptosis and Cancer
Editor: Albina W. Taylor, pp. 155-180

ISBN: 978-1-60021-506-3

Chapter IX

CATHEPSIN-REGULATED CELL DEATH PATHWAYS: LYSOSOMES AS TARGETS FOR CANCER THERAPY

Nagathihalli S. Nagaraj[1] and Wolfgang Zacharias[2]

[1]Department of Medicine, James Graham Brown Cancer Center, [2]Department of Medicine, James Graham Brown Cancer Center, and Department of Pharmacology & Toxicology, University of Louisville School of Medicine, Louisville, Kentucky 40202, USA.

ABSTRACT

Programmed cell death or apoptosis is a controlled physiological process that has become a focus for cancer therapy. The caspase cascade system plays vital roles in the initiation, transduction and amplification of intracellular apoptotic signals. Caspases exist as inactive zymogens in cells and undergo a cascade of catalytic activation at the onset of apoptosis. However, new evidence points to an emergent role of alternative caspase-independent cell death pathways. Many tumor cells can unexpectedly survive the activation of caspases; thus, caspase-independent cell death programs are gaining increasing interest in cancer research. Tumor invasion and metastasis are associated with altered lysosomal trafficking and increased expression of lysosomal cathepsin proteases. Lysosome stability and lysosomal cathepsins have recently been suggested to play an important role in the early stages of apoptosis induced by various agents. During lysosomal rupture, cathepsin B and D enzymes are released from the lysosomal compartment to the cytosol where they contribute to the apoptotic cascade upstream of mitochondria. Such role of lysosomes and lysosomal cathepsins in initiation and execution of the apoptotic program has become clear in several models, and the existence of a lysosomal pathway of apoptosis is now generally accepted. We have explored the role of cathepsins as upstream signals contributing to apoptotic cell death. Biochemical and mechanistic studies have led to important insights in the underlying molecular mechanisms of cathepsin regulation. Here, we highlight recent advancements in understanding the molecular mechanism of cathepsin-mediated cell death, and outline possible implications for the development of novel cancer therapeutics that target tumor cell lysosomes.

1. Introduction

Apoptosis is a genetically controlled event that is important in tissue development, homeostasis and disease. It is defined by a pattern of molecular and morphological changes that result in the 'packaging' and rapid removal of the dying cells (Chipuk and Green, 2005). While the importance of caspases in the apoptotic process is firmly established, the role of lysosome and particularly of lysosomal enzymes in cell death has only recently been unmasked.

The term "cathepsins", which was introduced in 1920 (Willstatter and Bamann, 1929), stands for "lysosomal proteolytic enzyme" regardless of the enzyme class. Therefore, in addition to cysteine proteases (cathepsins B, C, H, L, and S), this term also includes serine proteases (cathepsins A and G) and aspartic proteases (cathepsins D and E). Sequence homologies of the cysteine proteases indicate that these enzymes diverged early during eukaryotic evolution (Berti and Storer, 1995). Some of the genes encoding novel enzymes, *e.g.* cathepsins F, K, W, and X, may be the result of relatively recent gene duplication events, as suggested by three gene pairs with common chromosomal localization and high sequence homologies (Wex *et al.,* 1999; Chwieralski *et al.,* 2006). They are optimally active in the acidic pH of lysosomes (pH 4-5) but can also function in neutral pH outside the lysosomes with lower efficacies. The processing of the pro-cathepsins into the catalytically active form usually occurs within the lysosome (Ishidoh and Kominami, 2002); therefore, the enzymes escaping from the lysosomes are in their active monomeric form and generally do not require any other conformational change. Emerging evidence indicates that cathepsins have defined functions outside the lysosomal compartment, like degradation of extracellular matrix when secreted to the extracellular space and execution of programmed cell death (PCD) when released to the cytosol (Koblinski *et al.,* 2000; Jaattela, 2004). A role of lysosomal cathepsins in cell death had been first proposed in hamster kidney cells treated with a lysosomotropic detergent that led to cell death (Miller *et al.,* 1983). This observation was corroborated by the finding that photo-oxidative disruption of lysosomal membrane caused apoptosis of cultured human fibroblasts (Brunk *et al.,* 1997). A number of recent in vitro and in vivo studies have indicated that several lysosomal cathepsins can mediate or modulate cell death programs, including in cancer cell lines (Tardy *et al.,* 2006; Chwieralski *et al.,* 2006).

The activity of lysosomal cysteine cathepsins can be regulated in several ways, the most important being zymogen activation (Turk *et al.,* 2000) and inhibition by endogenous inhibitors, the cystatins (Turk and Bode, 1991; Turk *et al.,* 2002; Vigneswaran *et al.,* 2005). The cystatins constitute a superfamily of homologous proteins, which are subdivided into three families: stefins (family 1), cystatins (family 2) and kininogens (family 3) (Turk and Bode, 1991). The major role of cystatins is to protect the organism against endogenous proteases released from lysosomes, and against invading microorganisms and parasites that use cysteine proteases to enter the body (Turk *et al.,* 1997; Turk *et al.,* 2002). The importance of cystatins in preventing PCD under physiological and pathological conditions is demonstrated in cystatin B-deficient mice, which show increased apoptosis of cerebellar granule cells (Levine and Yuan, 2005). The forced expression of cystatin A also confers protection against lysosomal cell death pathways in hepatoma and fibrosarcoma cells (von Figura and Hasilik, 1986; Mortimore and Kadowaki, 1994), and high expression of cystatins in oral carcinoma cells confer a metastatic phenotype by enhancing their resistance to TNF-related apoptosis inducing ligand (TRAIL) induced apoptosis (Vigneswaran *et al.,* 2005). Cathepsin D is a lysosomal protease without known endogenous inhibitors. It is catalytically active at acidic pH values that vary according to substrates, and is mostly found in intracellular vesicles, lysosomes, phagosomes and late endosomes (Rochefort *et al.,* 2000).

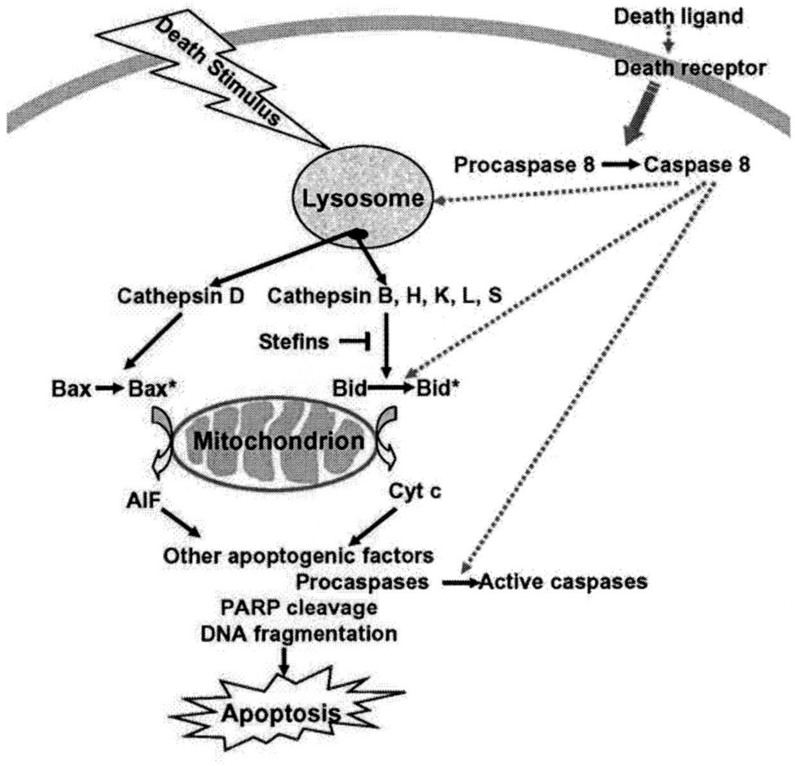

Figure 1. Model for lysosomal cathepsins-mediated apoptotic signaling pathways. Two initiation pathways, triggered by separate events, converge and execute apoptosis through the involvement of lysosomal cathepsins. Lysosomal membrane permeabilization (LMP) is triggered by direct or indirect (death ligand) stimulus, which induces the release of cathepsins to the cytosol. The released cathepsins cleave Bid, thereby inducing cytochrome c (cyt.c) release, followed by activation of effector caspases and resulting in apoptosis (Intrinsic pathway: dark line). On the other hand, TNFα, TRAIL or FasL (death ligand)-induced apoptosis activates caspase-8, thus further inducing LMP, Bid cleavage and/or direct caspase-3 activation (Extrinsic pathway: dotted line). Pro- and anti-apoptotic members of the Bcl-2 family (Bax and Bcl-2 as examples) regulate the stimulus response threshold. The pro-apoptotic Bid is at an intersection between the two pathways. The intrinsic and extrinsic pathways converge at the level of executioner caspases, thus leading to apoptosis. On the other hand, cathepsin D activated Bax, inducing apoptosis-inducing factor (AIF) release. In both pathways, Bak and/or Bax are responsible for the mitochondrial membrane permeabilization. Different pathways can be regulated by stefins (cysteine cathepsin inhibition). Additional details are given in the text.

Two types of PCD have been described: type I or apoptosis is characterized by a progressive retraction of chromatin and cytoplasm, followed by DNA internucleosomal fragmentation; type II or autophagic cell death is defined by accumulation of autophagic vacuoles (Tardy *et al.,* 2006). Apoptosis in response to cancer therapy proceeds through activation of the core apoptotic machinery including the extrinsic cell death receptor and the intrinsic mitochondrial signaling pathway (Herr and Debatin, 2001) (Figure 1). The intrinsic pathway is characterized by mitochondrial dysfunction with release of caspase activators, including cytochrome c, followed by activation of caspase-9 and caspase-3 (Hengartner, 2000). The extrinsic apoptotic pathway is initiated at the cell surface through death receptor-mediated activation of caspase-8 or caspase-10, followed by caspase-3 activation (Hengartner, 2000; Nagaraj *et al.,* 2004). This pathway may or may not be amplified by additional caspase-9/-3 activation via the intrinsic pathway (Krammer, 2000). Upon triggering of either pathway, caspase-3, -6 and -7, the final executioners of apoptosis, are activated, causing degradation of cellular proteins and leading to typical morphological changes such as chromatin condensation, nuclear shrinkage, and the formation of apoptotic bodies (Borner, 2003). Although caspase-3, -6 and -7 are all considered executioner caspases, the caspase-6 and -7 have minimal impact on this process. Interestingly, the link between extrinsic and intrinsic apoptosis occurs when caspase-8 induces the proteolytic cleavage of Bid to a truncated Bid (tBid) (Figure 1). tBid is capable of translocating from the cytosol to mitochondrial membranes where it stimulates more efficient oligomerization of Bax and activates the intrinsic pathway (Li *et al.,* 1998; Desagher *et al.,* 1999; Eskes *et al.,* 2000). However, the absolute requirement for caspase activation in apoptosis is no longer considered a dogma. During recent years it has become clear that cathepsins can also be involved in the regulation of apoptosis (Yamashima, 2000). Here, we discuss the morphological and biochemical phenotypes of lysosomal cathepsin-dependent cell death, present model systems that allow studies on cathepsin-dependent cell death, and propose mechanisms for lysosomal involvement in cell death and targeting lysosomes for future cancer therapy.

2. Cathepsin-Mediated Cell Death Pathways

Several members of the cathepsin family have been implicated in cancer progression, in particular the cysteine proteases cathepsins B and L and the aspartate protease cathepsin D (Jaattela, 2004; Guicciardi *et al.,* 2004). High expression levels of these cathepsins offer a reliable diagnostic marker for poor prognosis (Kos and Lah, 1998). Together with matrix metalloproteases and the plasminogen activator system, secreted cathepsins participate in the degradation of extracellular matrix, thereby enabling enhanced cellular motility, invasion, and angiogenesis. A stimulus that, directly or indirectly, induces lysosomal membrane permeabilization (LMP) will release the cysteine proteases cathepsin B, H, L, S, K and C (Li *et al.,* 2000; Guicciardi *et al.,* 2000; Reiners *et al.,* 2002; Cirman *et al.,* 2004; Nagaraj *et al.,* unpublished data), and the aspartic protease cathepsin D (Kagedal *et al.,* 2001a; Bidère *et al.,* 2003). Release of cathepsins from the lysosomes into the cytoplasm is a precondition for their participation in the regulation of apoptosis. Different mechanisms have been described which can contribute, probably in a stimulus- and cell-type-dependent fashion, to LMP (Ferri and

Kroemer, 2001; Brunk *et al.,* 2001). Once in the cytosol, the signaling pathway is cathepsin-specific, based on their capability to cleave and/or activate cytosolic factors such as Bid or Bax. Bid is cleaved and translocated to the mitochondria following lysosomal disruption by lysosomotropic agents (Cirman *et al.,* 2004). The cleavage site of lysosomal enzymes within Bid has been mapped at Arg65, a different site compared to the one cleaved by caspase-8 (Asp59) or granzyme B (Asp75), but located in the same flexible loop (Stoka *et al.,* 2001). Further studies confirmed that several purified cathepsins, such as B, K, L, H, and S, but not cathepsin D, C, and X, cleave Bid at Arg65, with the exception of cathepsin H, which has a unique cleavage site at Arg71 (Guicciardi *et al.,* 2004; Cirman *et al.,* 2004). However, proteolytic cleavage at Arg71 is not observed with lysosomal extracts, thus a role of cathepsin H in Bid cleavage in vivo seems unlikely (Stoka *et al.,* 2001; Guicciardi *et al.,* 2004; Cirman *et al.,* 2004).

On the other hand, cathepsin D activates the pro-apoptotic Bcl-2 homolog Bax, inducing selective release of the mitochondrial apoptosis inducing factor (AIF) and subsequent apoptosis (Bidère *et al.,* 2003). Release of cytochrome *c* from the mitochondrial intramembrane space to the cytoplasm is commonly mediated by the pro-apoptotic Bcl-2 family proteins Bak and Bax, which in apoptotic cells either oligomerize and form pores in the mitochondrial outer membrane (Desagher and Martinou, 2000). The requirement of Bax and/or Bak in lysosome-mediated apoptosis has been confirmed by demonstrating that embryonic fibroblasts from Bax/Bak double knock-out mice are resistant against mitochondrial membrane permeabilization induced by the lysosomotropic agents ciprofloxacin, norfloxacin, and hydroxychloroquine (Boya *et al.,* 2003a and 2003b). However, the absence of Bax and Bak in the double knock-out cells does not impair LMP, suggesting that lysosomal enzymes are not able to directly induce mitochondria dysfunction, but require the activation of a Bax/Bak-dependent pathway (Boya *et al.,* 2003b; Guicciardi *et al.,* 2004). It is thus clear that there is no generally conserved pathway of lysosome-mediated apoptosis; rather, the pathways differ, being stimulus- and cell type-dependent. Some models of apoptosis appear to be dependent exclusively on cathepsins (Foghsgaard *et al.,* 2001; Leist and Jaattela, 2001; Boya *et al.,* 2003a), whereas some models rely on both caspases and cathepsins for their initiation and execution (Faubion *et al.,* 1999; Guicciardi *et al.,* 2000). However, a growing body of evidence suggests that lysosomal cathepsins promote apoptosis by acting on mitochondria to induce mitochondrial dysfunction, associated with release of pro-apoptotic factors (Guicciardi *et al.,* 2000; Bidere *et al.,* 2003; Boya *et al.,* 2003a, b; Zhao *et al.,* 2003; Cirman *et al.,* 2004). Cathepsins are involved in cell death pathways triggered by many cytotoxic agents (Table 1). The release of lysosomal cathepsins may cause mitochondrial dysfunction directly (Zhao *et al.,* 2003) or indirectly (Guicciardi *et al.,* 2000; Stoka *et al.,* 2001; Bidere *et al.,* 2003; Boya *et al.,* 2003a and 2003b; Cirman *et al.,* 2004).

Cathepsin B

Cathepsin B, one of the most stable proteases at physiologic pH, is an essential mediator in TNFα-induced apoptosis in murine hepatocytes (Guicciardi *et al.,* 2000), tumor cells (Foghsgaard *et al.,* 2001; Mathiasen *et al.,* 2001; Taha *et al.,* 2005; Nagaraj *et al.,* 2006), immortalized murine embryonic fibroblasts (Fehrenbacher *et al.*, 2004), after brain ischemia (Yamashima *et al.*, 1998; Tsuchiya *et al.*, 1999), and in PC12 cell apoptosis after serum deprivation (Shibata *et al.*, 1998). It was proposed that cathepsin B contributes to apoptosis induced by DNA damage or TNF-α in tumor cells by causing loss of sphingosine kinase-1, a key enzyme that generates the anti-apoptotic lipid sphingosine 1-phosphate (Taha *et al.*, 2004 and 2005). Cathepsin B is also implicated in hepatocyte and tumor cell death induced by various insults such as bile salts (Roberts *et al.*, 1997; Faubion *et al.*, 1999), cold ischemia–warm reperfusion (Baskin-Bey *et al.*, 2005), low-frequency electromagnetic stimulation (Fedorowski *et al.*, 2004) or microtubule stabilizing agents (Broker *et al.*, 2004) (Table 1).

Table 1. Summary of cathepsin-dependent cell death events triggered by cytotoxic agents.

Cytotoxic agent	System	Cathepsin mediator during cell death	References
Camptothecin	Hepatocytes	Cathepsin D	Roberts *et al.*, 1999
Paclitaxel	Non-small cell lung cancer cells	Cathepsin B	Bröker *et al.*, 2004
Staurosporine	Fibroblasts	Cathepsin D	Johansson *et al.*, 2003
	T lymphocytes	Cathepsin D	Bidere *et al.*, 2003
Quinolone antibiotics	Several human and mice cells	Cathepsin B and D	Boya *et al.*, 2003a
Etoposide	Myelodysplastic syndrome-derived cells	Cathepsin L	Hishita *et al.*, 2001
Naphthazarin	Cardiomyocytes	Cathepsin D	Ollinger 2000
Hydrogen peroxide	Alveolar type II cells	Cathepsin D	Yin *et al.*, 2005
Bleomycin	Alveolar type II cells	Cathepsin D	Li *et al.*, 2005
Silica	Alveolar macrophages	Cathepsin D	Thibodeau *et al.*, 2004
Glycochenodeoxycholate (Bile salt)	Hepatoma cells	Cathepsin B	Faubion *et al.*, 1999
Epothilone B	Non-small cell lung cancer cells	Cathepsin B	Broker *et al.*, 2004
Discodermolide	Non-small cell lung cancer cells	Cathepsin B	Broker *et al.*, 2004

Main effector caspases, such as pro-caspase-3, -6, -7, as well as pro-caspase-2, -9, -12, and -14, are poor substrates for various cathepsins in vitro, including cathepsin B, H, K, L, S, and, X (Vancompernolle *et al.*, 1998; Stoka *et al.*, 2001; Guicciardi *et al.*, 2004). Addition of purified cathepsin B or D to mitochondria in vitro results in substantial generation of reactive oxygen species (ROS) (Zhao *et al.*, 2003). In contrast, when isolated mitochondria are incubated with purified cathepsin B in the presence of cytosolic extracts, a substantial release of cytochrome c is detected, suggesting that cytosolic factor(s) activated by cathepsin B mediate mitochondrial dysfunction (Guicciardi *et al.*, 2000). Specific cathepsin B inhibitors, such as CA-074, CA-074Me or cystatin A, can protect cells against PCD (Jones *et al.*, 1998; Yamashima *et al.*, 1998; Tsuchiya *et al.*, 1999; Li *et al.*, 2005; Vigneswaran *et al.*, 2005; Nagaraj *et al.*, 2006). Ribozymes that target cathepsin B mRNA, when stably expressed in oral squamous cell carcinoma cell lines, reduce invasiveness and suggest the possibility to control oral carcinoma malignancy and metastasis by RNA inhibitors (Wickramasinghe *et al.*, 2005). Silencing of cystatin M by small interference RNAs (RNAi) in tumor cells not only increases their invasion and motility via a cysteine-proteinase-dependent pathway, but also renders them hyperproliferative through a currently unknown mechanism (Vigneswaran *et al.*, 2006). Conversely, RNAi-mediated inhibition of cathepsin B leads to decreased cell invasion, angiogenesis and tumor growth in gliomas (Gondi *et al.*, 2004). The role of cathepsin B in apoptosis appears to be dependent on context and/or cell type, possibly because of partial redundant actions of different cathepsins. Further discrepancies between the above observations may stem from two still obscure issues, the site of cathepsin action (intracellularly or extracellularly) and the nature of their substrates (Joyce and Hanahan, 2004). Therefore, cathepsin B may play two opposing roles in malignancy: as an executioner initiator of apoptosis in cytotoxic intracellular signaling cascades and as a mediator of tumor invasion when secreted.

Cathepsin L

Cathepsin L, the least stable of the lysosomal cysteine proteases at neutral pH, has also been implicated in the apoptotic process. It is an important regulator of ultraviolet-induced apoptosis of keratinocytes (Tobin *et al.*, 2002; Welss *et al.*, 2003), and etoposide-induced apoptosis of P39 cells (Hishita *et al.*, 2001). Human cathepsin L-like protease (Ishisaka *et al.*, 1999) and the trypanosomal cysteine cathepsin cruzipain (Stoka *et al.*, 2001), structurally related to cathepsin L, cleave procaspase-3/-7 in cell-free systems and cell culture. Cathepsin L has also been proposed to trigger activation of procaspase-3 indirectly via activation of another unidentified lysosomal, membrane-bound protease (Katunuma *et al.*, 2001). Murine cathepsin L-deficient immortalized embryonic fibroblasts were resistant to TNF-α-induced cell death (Fehrenbacher *et al.*, 2004), and involvement of this protease in β-amyloid-induced apoptosis in rat cortical neurons (Bol and Campbell, 2004) has been reported, based on the inhibition of cell death by pharmacological inhibitors of cathepsin L. However, expression of cathepsin L antisense was shown to increase staurosporine-induced apoptosis in glioma cell lines, whereas overexpression of this enzyme resulted in inhibition of cell death (Levicar *et al.*, 2003). The pro-apoptotic role of cathepsin L thus remains controversial and needs to be clarified further.

Cathepsin D

Many reports have implicated the lysosomal aspartyl protease cathepsin D in cell death (Tardy *et al.*, 2006). This cathepsin is expressed in most mammalian tissues (Barrett and Kirschke, 1981); however, its expression varies considerably depending on the cell type

(Reid *et al.,* 1986). Cathepsin D represents the majority of the endopeptidase activity within lysosomes. Its structure is very similar to that of other aspartyl proteases of the pepsin family (Tang and Wong, 1987) but, contrary to those that contain only one peptidic chain, cathepsin D is formed by two subunits of 48 and 32 kDa (Hasilik and Neufeld, 1980; Tardy *et al.,* 2006). It has been implicated in apoptosis induced by staurosporine (Bidere *et al.,* 2003; Johansson *et al.,* 2003), interferon- , Fas/CD95/APO-1 and TNF-α (Deiss *et al.,* 1996; Demoz *et al.,* 2002), oxidative stress (Roberg and Ollinger, 1998; Roberg *et al.,* 1999; Ollinger, 2000; Kagedal *et al.,* 2001a), sphingosine (Kagedal *et al.,* 2001b), and p53 (Wu *et al.,* 1998) (Table 1). PC12 cells overexpressing cathepsin D died more rapidly than wild-type cells (Shibata *et al.,* 1998). Moreover, microinjection of this enzyme into fibroblasts induced caspase-dependent apoptosis (Roberg *et al.,* 2002).

RNAi-mediated silencing of cathepsin D expression protected leukemic cells against TNF-α, TRAIL or chemotherapeutic drugs (Emert-Sedlak *et al.,* 2005). The aspartic protease inhibitor pepstatin A blocked cell death induced by various stress factors, including Fas, IFN-γ (Deiss *et al.,* 1996), TNF-α (Demoz *et al.,* 2002; Heinrich *et al.,* 2004), camptothecin (Roberts *et al.,* 1999), naphthazarin (Roberg and Ollinger, 1998; Ollinger, 2000), hydrogen peroxide (Yin *et al.,* 2005), staurosporine (Johansson *et al.,* 2003), the synthetic retinoid CD437 (Zang *et al.,* 2001), sphingosine (Kagedal *et al.,* 2001b), bleomycin (Li *et al.,* 2004), silica (Thibodeau *et al.,* 2004), and the induction of wild type p53 in a murine lymphoma cell line (Wu *et al.,* 1998). On the other hand, inhibition of cathepsin D by pepstatin A in murine hepatoma cells did not suppress Bid cleavage nor pro-caspase-9 and -3 activation mediated by phototherapy (Reiners Jr *et al.,* 2002). Moreover, stress-induced apoptosis was impaired neither in bovine cathepsin D-deficient fibroblasts nor in pepstatin A-treated human fibroblasts or lymphoblasts (Tardy *et al.,* 2003). In addition, pepstatin A did not protect PC12 cells from cell death induced by serum deprivation (Isahara *et al.,* 1999) nor L929 fibroblasts from etoposide-induced apoptosis (Demoz *et al.,* 2002), and triggered apoptosis in human neuroblastoma cell lines (Castino *et al.,* 2002). Therefore, the role of cathepsin D in PCD appears as a controversial issue, again emphasizing some cell type- and/or cell death stimulus-specificity, and implying variable target substrate(s) once the protease is released from the lysosome.

Thus, consistent with their relatively low substrate specificities, lysosomal cathepsins trigger apoptosis not via a single pathway, but rather via multiple molecular pathways, which often merge with those controlled by traditional mediators of apoptosis, like caspases and Bcl-2 family proteins (Figure 1). As the key role of lysosomes in apoptosis is becoming increasingly evident, other molecular pathways mediated by lysosomal cathepsins are likely to emerge.

3. Cathepsins as Potential Targets for Tumor Cell Killing

In cancer cells, especially those at the invasive edges of tumors, the localization of lysosomes is often altered from perinuclear to peripheral regions and the lysosomal content is secreted into the extracellular space (Koblinski *et al.,* 2000). In the context of cancer therapy, it is intriguing that cathepsins are able to mediate caspase- and mitochondrion-independent PCD, allowing cell death to occur even in cells with multiple defects in the classic apoptosis pathway characterized by chromatin condensation (Jaattela, 2004). An increase in expression, secretion and/or activity of lysosomal cathepsins (primarily cathepsins B, D and L) has been demonstrated in a variety of human tumors, including breast, lung, oral and brain cancers (Vigneswaran *et al.,* 2000; Berchem *et al.,* 2002). Apparently, enhanced secretion of these proteases by exocytosis promotes cancer growth, invasion, and metastasis by promoting degradation of extracellular connective matrices and supporting angiogenesis (Koblinski *et al.,* 2000). A correlation between expression levels of cathepsins and aggressiveness of breast, prostate, oral or gastric cancers has been proposed (Visscher *et al.,* 1993; Bremer *et al.,* 2002; Vigneswaran *et al.,* 2000; Kielan *et al.,* 2004; Nomura and Katunuma, 2005), and the diagnostic and prognostic potential of lysosomal cathepsins as cancer markers has been evaluated (Schwartz, 1995; Kirschke, 1997; Kos *et al.,* 2000; Rochefort *et al.,* 2000; Berdowska, 2004). In some types of tumors, the high content of lysosomal cathepsins correlated with a higher risk of recurrence and with a poor prognosis (Thomssen *et al.,* 1995; Losch *et al.,* 2004). Indeed, inhibition of cathepsin B by synthetic cysteine protease inhibitors effectively reduced the invasiveness of cancer cell lines (Demchik *et al.,* 1999; Li *et al.,* 2005).

Tumor cell death can be caspase-independent due to acquired defects in the classic caspase-dependent pathways of apoptosis during the malignant transformation (Foghsgaard *et al.,* 2001; Leist and Jaattela, 2001), and such alternative pathways may represent potential targets for cancer therapy (Jaattela, 2002). Enhancing the cathepsin-mediated cell death pathway may be a therapeutic strategy to overcome deficiencies in caspase-dependent cell death. The overexpression of cathepsins often observed in tumors should render cancer cells more susceptible to cell death via the lysosomal pathway. It should be possible to exploit this pathway to differentially modulate susceptibility of cancer cells versus non-malignant cells to cell death. TRAIL preferentially kills tumor cells both in vitro and in vivo (Ashkenazi *et al.,* 1999; Walczak *et al.,* 1999). Although the underlying mechanism of TRAIL selectivity toward tumor cells remains unclear, TRAIL can trigger lysosomal cell death (Foghsgaard *et al.,* 2001; Nagaraj *et al.,* 2006) and may induce apoptosis in transformed cells via the lysosomal pathway.

Whereas extracellular cathepsin inhibitors might function as tumor suppressors through their ability to inhibit the tumor-promoting actions of cathepsins outside the cell, intracellular inhibitors can rescue cells from the consequences of partial LMP and thereby promote tumor cell survival (Kroemer and Jaattela, 2005). No endogenous inhibitors of cathepsin D are known, but cysteine cathepsins can be inhibited by at least four cytosolic protease inhibitors - cystatin A, cystatin B, squamous cell carcinoma antigen 1 (SCCA1, also known as serpin B3) and serpin a3g (Kang *et al.,* 1999; Kroemer and Jaattela, 2005). The potential relevance of endogenous cathepsin inhibitors to cancer malignancy is indicated by the negative prognostic value of cystatin A and SCCA1 expression in breast cancer and squamous cell carcinomas, respectively (Kuma *et al.,* 2004; Kroemer and Jaattela, 2005), the increased tumorigenic potential of SCCA1-expressing tumor cells (Lum *et al.,* 2005), and the ability of NF-κB to induce serpin a3g (Kang *et al.,* 1999).

Successful treatment with certain chemotherapeutic drugs is dependent on their ability to trigger cell death in tumor cells, and activation of apoptosis is at least partially involved in this process (Fisher, 1994). The majority of cytotoxic agents trigger the mitochondria pathway, but death receptors are also involved in chemotherapy-induced apoptosis (Friesen *et al.,* 1999; Green, 2000). However, recent evidence suggests that there are forms of chemotherapy-induced cell death that cannot readily be classified as apoptosis or necrosis but fit more in the apoptosis-like/necrosis-like PCD model (Leist and Jäättela, 2001; Schmitt and Lowe, 2001). While there is need for a better understanding of the participation of cathepsins in cell death, whether apoptotic or autophagic, lysosomal proteins appear to be promising potential therapeutic targets for the development of novel strategies to treat malignancies (Jaattela, 2002; Kroemer and Jaattela, 2005; Fehrenbacher and Jaattela, 2005). Efforts should be made to antagonize the effects of secreted cathepsins that promote invasiveness, mitogenesis, and/or anti-tumoral immune response of tumor cells, while stimulating or restoring the intracellular machinery that uses lysosomal constituents to enhance cancer cell death (Erdal *et al.,* 2005).

The activation of cathepsin-dependent death pathways could prove useful in cancer therapy. Conversely, tumors that manage to suppress such alternative death programs might have a selective growth advantage. Indeed, tumor progression is associated with changes that confer resistance to the cysteine cathepsin-mediated death pathway, *i.e.*, activation of NF-κB that induces the expression of potent cytosolic serine and cysteine protease inhibitor SpiA2 (Liu *et al.,* 2003), increased activity of phosphoinositol-3'-kinase (PI3K) that protects cells from cathepsin-mediated death (Madge *et al.,* 2003), or increased expression of endogenous cysteine cathepsin inhibitors (Kuopio *et al.,* 1998). Also, depletion of heat shock protein Hsp70 in cancer cells results in a cysteine cathepsin-dependent cell death, which is preceded by lysosomal destabilization and release of lysosomal constituents to the cytosol (Nylandsted *et al.,* 2004).

4. Lysosomal Control of Cell Death

Lysosomes, which have been given the epithet 'suicide bags', contain a large number of catabolic hydrolases with an acidic pH optimum. They contribute to type II (autophagic) cell death, although the exact molecular mechanisms of this type of cell removal have not been elucidated (Ferri and Kroemer, 2001). Lysosomes undergo membrane permeabilization in response to extra- or intra-lysosomal stimuli (Turk *et al.,* 2002; Guicciardi *et al.,* 2004). Direct stimuli may involve lysosomotropic agents (Li *et al.,* 2000; Cirman *et al.,* 2004), photodamage (Reiners *et al.,* 2002), sphingosine (Kagedal *et al.,* 2001b), D-galactosamine (Katunuma *et al.,* 2004), ROS (Liu *et al.,* 2004) and compounds that activate p53-independent apoptosis (Erdal *et al.,* 2005), whereas indirect stimuli may involve caspase-8 activation following binding of ligands like TNF-α or TRAIL to their receptors (Guicciardi *et al.,* 2000; Foghsgaard *et al.,* 2001; Liu *et al.,* 2003; Tardy *et al.,* 2004; Nagaraj *et al.,* 2006) (Figure 1). Lysosome-mediated apoptosis can be either death receptor-dependent (Guicciardi *et al.,* 2000; Foghsgaard *et al.,* 2001; Liu *et al.,* 2003; Tardy *et al.,* 2004) or -independent (Stoka *et al.,* 2001; Reiners *et al.,* 2002; Cirman *et al.,* 2004) (Figure 1). LMP and cathepsins also participate in type I cell death induced by classic apoptotic stimuli such as death receptor or p53 activation, as well as by drugs that were not primarily designed to target lysosomes (von Figura and Hasilik, 1986; Stein *et al.,* 1987; Lobel *et al.,* 1989) (Figure 2).

Sphingosine induces lysosomal dysfunction in primary mouse hepatocytes and permeabilizes isolated hepatic lysosomes in vitro (Werneburg *et al.,* 2002). Ceramide generated by the endolysosomal acid sphingomyelinase binds to and activates cathepsin D (Werneburg *et al.,* 2002). ROS can induce lysosomal leakage, possibly as a consequence of intralysosomal iron-catalyzed oxidative processes (Antunes *et al.,* 2001; Persson *et al.,* 2003), and ROS-induced LMP usually precedes mitochondrial dysfunction (Roberg and Ollinger, 1998; Roberg *et al.,* 1999; Guicciardi *et al.,* 2000). However, lysosomal enzymes have been found to act on mitochondria and promote mitochondrial ROS generation, thereby creating a feedback loop that can lead to more LMP (Zhao *et al.,* 2003). Recent studies have shown that the phosphorylated, active form of Bcl-2 blocks oxidant-induced apoptosis, at least in part, by stabilizing lysosomes (Zhao *et al.,* 2001). In a model for TNF-α induced hepatocyte apoptosis, cathepsin B knock-out in these cells confers increased resistance to LMP, suggesting that lysosomal enzymes are not only passively released but also participate in membrane destabilization from within the lysosome (Werneburg *et al.,* 2002). These findings are supported by the evidence that serine protease inhibitor 2A (Spi2A), which also effectively inhibits cathepsin B, also partially reduces lysosomal breakdown (Liu *et al.,* 2003). An increase in lysosomal size often observed during the early stages of apoptosis may determine their susceptibility to rupture; indeed, larger lysosomes may breakdown more efficiently with stress (Ono *et al.,* 2003).

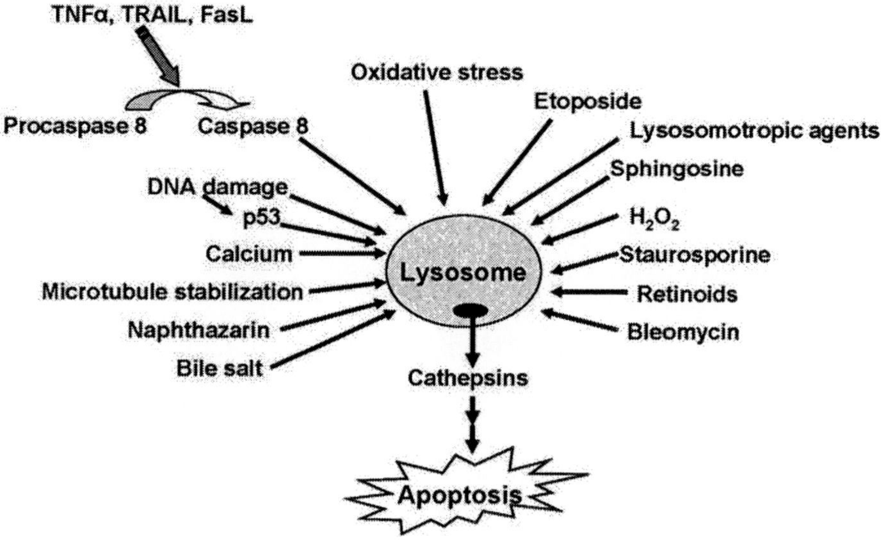

Figure 2. Model for the role of lysosomes in various forms of cell death stimulus. Different types of cell death stimuli trigger lysosomal membrane permeabilization (LMP) to release cathepsins. Lysosomes can be accessed via endocytosis, pinocytosys or via membrane diffusion. Depending on the cell type, stimulus, and extent of LMP, the resulting release of lysosomal cathepsins leads to the activation of pro-apoptotic factors and thus apoptosis. Additional details are given in the text.

The release of cathepsins from lysosomes is a size-selective process that does not necessarily lead to destruction and de-acidification of lysosomes (Bidere *et al.,* 2003; Chwieralski *et al.,* 2006). In addition to extralysosomal signals, processes within the lysosomes are important to promote LMP, and the magnitude of LMP is associated with the type of cell death (Li *et al.,* 2000; Kagedal *et al.,* 2001b). For example, the stress-inducible heat shock protein Hsp70 is an integral part of the lysosomal membrane and has important anti-apoptotic functions. Depletion of Hsp70 induces the release of lysosomal enzymes into the cytosol, leading to caspase-independent apoptosis (Chwieralski *et al.,* 2006). Such involvement of lysosomes and lysosomal cathepsins in apoptosis has been confirmed in several systems (Guicciardi *et al.,* 2000; Foghsgaard *et al.,* 2001; Stoka *et al.,* 2001; Reiners *et al.,* 2002; Boya *et al.,* 2003a and 2003b; Cirman *et al.,* 2004; Yin *et al.,* 2005).

In cells undergoing apoptosis, leakage of lysosomal cathepsins must be finely controlled since there is no complete breakdown of the lysosomal membrane, as shown by the fact that lysosomes are still intact and able to retain the pH gradient (Foghsgaard *et al.,* 2001; De′moz *et al.,* 2002). On the other hand, massive leakage of lysosomal cathepsins is expected to induce necrotic cell death rather than apoptosis (Bursch, 2001). It should be noted that lysosomotropic agents (Li *et al.,* 2000) and the synthetic retinoid CD437 (Zang *et al.,* 2001) also cause leakage of lysosomal cathepsins associated with induction of apoptosis. How this leakage occurs is not known, though it is probably favored by the formation of free radicals in the lumen of lysosomes via a Fenton-like reaction (Roberg *et al.,* 1999; Brunk and Svensson, 1999; Ollinger, 2000). The lysosomal–autophagic system is probably also involved in apoptosis, perhaps only in the initial steps. Thus, the lysosome acts as a central organelle that determines which type of death pathway will dominate in injured cells as a result of the type, intensity and duration of the cytotoxic insult (Ferri and Kroemer, 2001).

Clearly, in order to participate to the apoptotic process, lysosomal enzymes need to be translocated to the cytosol (possibly to the nucleus) where most of the cellular proteins degraded during apoptosis are found (Bidere *et al.,* 2003). However, the mechanism(s) leading to partial or selective LMP, to the translocation of only a moderate amount of lysosomal enzyme to the cytosol, and the engagement of death receptors in apoptosis associated with lysosomal dysfunction still need to be clarified.

5. Lysosome-Mediated Cell Death Pathways in Cancer Cells

A number of observations on cancer cells have shown that cathepsins can translocate from the lysosome lumen to the cytosol in response to a variety of death stimuli such as TNF-α (Foghsgaard *et al.,* 2001; Mathiasen *et al.,* 2001), retinoids (Zang *et al.,* 2001), etoposide (Emert-Sedlak *et al.,* 2005), Fas, oxidative stress, and growth factor starvation (Brunk and Svensson, 1999) (Figure 2). Given the potentially fatal outcome of LMP, it is not surprising that cancer cells have developed strategies to counteract LMP and its consequences (Majeski and Dice, 2004). One of the mechanisms, which seems to be confined exclusively to cancer cells, is the translocation of Hsp70 to the inner leaflet of the endo-lysosomal membranes (Koster *et al.,* 1993). Hsp70 is abundantly expressed in malignant tumors of various origins

(Jäättelä, 1999), and its expression correlates with increased cell proliferation, poor differentiation, lymph node metastases, and poor therapeutic outcome in human breast cancer (Lazaris *et al.,* 1997; Vargas-Roig *et al.,* 1998). Furthermore, numerous experimental cancer models have demonstrated the tumorigenic potential of Hsp70 in rodents (Volloch and Sherman, 1999; Gurbuxani *et al.,* 2001). Hsp70 apparently promotes tumorigenesis via its pro-survival function, since it can effectively inhibit cell death induced by a wide range of stimuli including cancer-related stress factors like hypoxia, inflammatory cytokines, monocytes, irradiation, oxidative stress, and anticancer drugs (Simon *et al.,* 1995; Mosser *et al.,* 1997; Jäättelä *et al.,* 1998; Vayssier *et al.,* 1998; Nylandsted *et al.,* 2004). Conversely, the depletion of Hsp70 in human tumor cells specifically triggers LMP and cathepsin-dependent cell death in vitro and in tumor xenografts in vivo (Koster *et al.,* 1993; Bursch, 2001; Levine and Yuan, 2005). Hsp70 is often overexpressed in malignant human tumors, correlating with a poor prognosis in patients with breast and endometrial cancer (Lockshin and Zakeri, 2004). Expression of Hsp70 at the cell surface seems to be a strictly tumor-specific phenomenon, found in as many as 80% of freshly isolated human tumors of various origins (Baehrecke, 2005; Kroemer and Jäättelä, 2005).

PI3K is commonly activated in human cancer cells and may also function as an LMP inhibitor (Kroemer and Jäättelä, 2005). Among its many functions, PI3K controls the maturation, size, stability and activity of lysosomes (Halestrap and Gillespie, 2000; Debnath *et al.,* 2005). Inhibition of PI3K in human vascular endothelial cells induces the translocation of cathepsin B to the cytosol, arguing for a direct role of PI3K in preserving lysosomal membrane integrity (Debnath *et al.,* 2005). Furthermore, PI3K inhibitors sensitize endothelial cells to cytokine-induced lysosomal cell death pathways (Debnath *et al.,* 2005). However, the precise role of PI3K as an LMP inhibitor remains to be studied (Kroemer and Jäättelä, 2005).

Several mechanisms have been proposed that may connect the lysosomal pathway of cell death to other death pathways. It is still unknown whether the molecular signaling pathways leading to lysosomal rupture are the same in death receptor- and DNA damage-induced cell death (Tardy *et al.,* 2006). Lysosomal leakage may also represent a common mechanism for p53-dependent (Yuan *et al.,* 2002) and p53-independent (Erdal *et al.,* 2005) apoptosis. This latter observation has important implications in the development of anticancer therapies based on induction of lysosomal cell death since p53 is frequently mutated in human tumors (Tardy *et al.,* 2006). The possibility that lysosomal cathepsins themselves may permeabilize the lysosomal membrane cannot be ruled out (Tardy *et al.,* 2006). Indeed, lysosomal leakage is strongly decreased in cathepsin B-deficient cells after TNF-α treatment (Werneburg *et al.,* 2002). Moreover, the serine protease inhibitor Spi2A, which also inhibits cathepsin B, partially blocks lysosomal membrane leakage (Liu *et al.,* 2003), and NF-κB may inhibit the lysosomal pathway of apoptosis by increasing Spi2A transcription (Liu *et al.,* 2003).

Lysosomal rupture leads to cytosol acidification that can activate other factors which may participate in lysosomal PCD (Tardy *et al.,* 2006). During drug-induced apoptosis of tumor cells, the intracellular increase in H_2O_2 is responsible for early cytosolic acidification leading to Bax translocation into mitochondria (Hirpara *et al.,* 2001; Ahmad *et al.,* 2004). Accordingly, signals that inhibit cytosol acidification prevent human leukemia cell death (Ahmad *et al.,* 2004). Recently, it has been proposed that acidification could mediate a post-translational modification of an anti-apoptotic serine protease inhibitor, leukocyte elastase inhibitor (LEI) that converts it into lysosomal DNase II; this DNase could induce the apoptosis-typical nuclear DNA fragmentation (Altairac *et al.,* 2003; Tardy *et al.,* 2006).

Taken together, the control of the lysosomal cell death pathways during tumor progression resembles that of apoptosis in that transformation initially sensitizes cells to both death programmes, which then have to be disabled for malignancy to progress (de Duve 1983; Majeski and Dice, 2004).

6. Lysosomes as Targets for Future Cancer Therapy

Recent evidence argues that both classic apoptosis pathways and lysosomal death pathways must be suppressed for effective development and progression of cancer (Jaattela, 2004). Also, a newly emerging concept suggests that cathepsins have a dual role in tumor progression, as they increase malignancy by promoting invasion when outside the cell, but counteract it through their pro-apoptotic function when released into the cytosol (Kroemer and Jäättelä, 2005). Therefore, inhibiting the secretion of cathepsins could give rise to populations of lysosomes that are more prone to membrane rupture and thereby inhibit invasion and sensitize cells to the lysosomal cell death pathways (Kroemer and Jäättelä, 2005). Lysosomes in advanced-stage tumors are abnormal in their content, subcellular localization and function, which offers the possibility to target them for the specific eradication of tumors (Kroemer and Jaattela, 2005). The increased activity of PI3K, characteristic of many cancers, may contribute to the stability of tumor cell lysosomes (Fehrenbacher and Jäättelä, 2005). Among its many functions, PI3K controls the maturation, size, and activity of lysosomes (Mousavi *et al.,* 2003), and the inhibition of PI3K can shift the TNF-induced death pathway from caspase-dependent to cathepsin-dependent (Madge *et al.,* 2003).

Immortalization and oncogene-driven transformation sensitize cells to lysosomal cell death pathways, and many chemotherapeutic agents (like staurosporine, sphingosine, etoposide) trigger LMP in cancer cells (Kroemer and Jäättelä, 2005). However, it is unclear whether lysosomes in advanced tumors are actually more fragile than those in normal cells or whether additional changes during tumorigenesis overcome the initial destabilization of lysosomes. Expression of cathepsin inhibitors of the cystatin (Kuopio *et al.,* 1998) or serpin families (Liu *et al.,* 2003) may increase the viability of cancer cells with enhanced lysosomal activity. Tumor cells may also protect themselves against lysosomal leakage by translocation of Hsp70 into the lysosomal membrane, leading to more stable lysosomes (Nylandsted *et al.,* 2004). The membrane localization of Hsp70, found in 80% of human cancer biopsies, provides tumor lysosomes with increased resistance to LMP (Koster *et al.,* 1993; Baehrecke, 2005). On the other hand, depletion of Hsp70 triggers a tumor cell–specific lysosomal death program (Fehrenbacher and Jäättelä, 2005). As this subcellular localization of Hsp70 to the inner leaflet of the lysosomal membrane seems to be cancer-specific, targeting Hsp70 might have selective therapeutic effects. Introducing Hsp70 inhibitors that interfere with the membrane-binding of Hsp70 by the endocytic route may be an opportunity to interfere with its pro-survival and tumor-promoting function (Kroemer and Jäättelä, 2005). Thus, drugs that inhibit the PI3K pathway, the lysosomal localization of Hsp70, or the activity of cathepsin inhibitors are likely to sensitize tumor cells to anti-cancer drugs by facilitating LMP (Fehrenbacher and Jäättelä, 2005).

The low pH in the lysosomal compartment of tumor cells can be exploited bydesigning acid-labile pro-drugs that would be activated inside the lysosomes (Kroemer and Jäättelä, 2005). This can be achieved by incorporating acid-sensitive spacers between the drug and carrier that are specifically cleaved by lysosomal cathepsins (Tardy *et al.,* 2003). This approach could also be used to deliver compounds with direct membrane toxicity safely into the lysosomal lumen (Kroemer and Jäättelä, 2005). On the other hand, raising the lysosomal pH by pharmacological inhibition of the vacuolar proton pump increases the cytoplasmic retention of cisplatin, 5-fluorouracil and vinblastine, and ameliorates the nuclear targeting of doxorubicin, thereby sensitizing tumor cells to these treatments (Turk *et al.,* 2000). Such neutralization of lysosomal pH can also sensitize cancer cells to other treatments, possibly by inhibiting autophagy-mediated cytoprotection (Turk *et al.,* 1997), or by exerting direct cytotoxic effects through the induction of lysosomal dysfunction, LMP and cathepsin-mediated PCD (Guicciardi *et al.,* 2000; Foghsgaard *et al.,* 2001; Kroemer and Jäättelä, 2005).

A current major goal of research in cancer chemotherapy is the development of carrier systems for the delivery and site-specific uptake of the cytotoxic cargo in tumor tissue (Fehrenbacher and Jaattela, 2005). Since lysosomes can be reached easily via endocytosis, it seems worthwhile to focus on delivery systems that specifically target this organelle with appropriate drugs in order to overcome chemoresistance and obtain tumor regression. The therapeutic exploitation of the differences between normal and cancer cell lysosomes requires a better knowledge of the molecular changes that they experience during tumorigenesis. A thorough mass spectrometric analysis of the cancer-associated changes in lysosomal composition, like the MALDI-MS or electrospray ionization tandem MS used for the detection of enzymatic products of lysosomes, combined with systematic functional studies, may reveal novel targets for cancer therapy (Kroemer and Jäättelä, 2005). Also, more defined roles for each of the cathepsins need to be determined by dissecting their non-redundant functions in both tumor progression and cell death. Because different tumor cells and tissues exhibit rather specific characteristics in their increased cathepsin levels, altered lysosomal trafficking, and shifts in different endolysosomal populations, drug screens to identify molecules that induce LMP may prove effective in cancer treatment (Castino *et al.,* 2003; Fehrenbacher and Jaattela, 2005). Recent study on the screening of a small molecule drug library identified 175 compounds that induced death in cultured HCT116 colon cancer cells (Erdal *et al.,* 2005). Importantly, over half of the 11 compounds that induced significant cell death in p53-null cells triggered LMP and cathepsin-mediated killing of tumor cells (Erdal *et al.,* 2005).

A remaining key question is whether there are cancer-specific subpopulations of lysosomes that can account for lysosomal secretion and/or leakage in cancer cells (Kroemer and Jäättelä, 2005). The nature and spectrum of lysosomal proteins modulating cell death, their mode of activation and release, their target substrates, the regulatory pathways and crosstalk with other death-signaling pathways need to be resolved. Newly developing approaches such as metabolomics or comprehensive and quantitative proteomics techniques will expand our knowledge on the contribution of lysosomal proteins to cancer (Journet and Ferro, 2004). Without doubt, our recent progress in lysosome research warrants additional investment into the detailed elucidation of tumor-associated lysosomal death pathways and lysosomal changes, which will most likely lead to improvements in the fight against cancer.

7. Conclusion

Induction of apoptosis selectively in malignant cells has become a major focus for cancer therapy. New evidence points to an emerging role of alternative caspase-independent cell death pathways mediated by lysosomal cathepsin proteases. Lysosome stability and lysosomal cathepsins have recently been suggested to play an important role in the early stages of apoptosis induced by various agents. Thus, cathepsins appear to have dual and opposing roles in tumor progression by either functioning as invasion-supporting and thus tumor-promoting proteases, or as pro-apoptotic and thus tumor-inhibiting factors. Such functions of cathepsins presumably are dictated by their cellular localization, since the pro-apoptotic function requires cytosolic activity, whereas invasiveness is promoted by secretion of proteases. Diminished lysosome stability and membrane leakage, induced by either extracellular or internal apoptosis stimuli, will result in such elevated cytosolic cathepsin activities. Also, there is evidence to suggest lower stability of lysosomes in tumor cells compared to non-malignant cells. Thus, molecular factors that determine or alter lysosome stability and integrity may be novel and valuable targets for molecular cancer therapy, and lysosome-destabilizing drugs need to be considered for the development of novel cancer therapeutics that target tumor cell lysosomes.

References

Ahmad, K. A., Iskandar, K. B., Hirpara, J. L., Clement, M. V. & Pervaiz, S. (2004). Hydrogen peroxide-mediated cytosolic acidification is a signal for mitochondrial translocation of Bax during drug-induced apoptosis of tumor cells. *Cancer Res. 64*, 7867-7878.

Altairac, S., Zeggai, S., Perani, P., Courtois, Y. & Torriglia, A. (2001). Apoptosis induced by Na+/H+ antiport inhibition activates the LEI/L-DNase II pathway. *Cell Death Differ. 10*, 548-557.

Antunes, F., Cadenas, E. & Brunk, U. T. (2001). Apoptosis induced by exposure to a low steady-state concentration of H2O2 is a consequence of lysosomal rupture. *Biochem J. 356*, 549-555.

Ashkenazi, A., Pai, R. C., Fong, S., Leung, S., Lawrence, D. A., Marsters, S. A., Blackie, C., Chang, L., McMurtrey, A. E., Hebert, A., DeForge, L., Koumenis, I. L., Lewis, D., Harris, L., Bussiere, J., Koeppen, H., Shahrokh, Z. & Schwall, R. H. (1999). Safety and antitumor activity of recombinant soluble Apo2 ligand. *J Clin Invest. 104*, 155-162.

Baehrecke, E. H. (2005). Autophagy: dual roles in life and death? *Nat Rev Mol Cell Biol.* 6, 505-510.

Barrett, A. J. & Kirschke, H. (1981). Cathepsin B, cathepsin H, and cathepsin L. *Methods Enzymol. 80*, 535-561.

Baskin-Bey, E. S., Canbay, A., Bronk, S. F., Werneburg, N., Guicciardi, M. E., Nyberg, S. L. & Gores, G. J. (2005). Cathepsin B inactivation attenuates hepatocyte apoptosis and liver damage in steatotic livers after cold ischemia–warm reperfusion injury. *Am J Physiol.: Gastrointest Liver Physiol. 288*, G396-G402.

Berchem, G., Glondu, M., Gleizes, M., Brouillet, J. P., Vignon, F., Garcia, M. & Liaudet-Coopman, E. (2002). Cathepsin-D affects multiple tumor progression steps in vivo: proliferation, angiogenesis and apoptosis. *Oncogene, 21*, 5951-5955.

Berdowska, I. (2004). Cysteine proteases as disease markers. *Clin Chim Acta, 342*, 41–69.

Berti, P. J. & Storer, A. C. (1995). Alingment /Phylogeny of the papain superfamily of cysteine proteases. *J Mol Biol. 246*, 273– 283.

Bidere, N., Lorenzo, H. K., Carmona, S., Laforge, M., Harper, F., Dumont, C. & Senik, A. (2003). Cathepsin D triggers Bax activation, resulting in selective apoptosis-inducing factor (AIF) relocation in T lymphocytes entering the early commitment phase to apoptosis. *J Biol Chem. 278*, 31401–31411.

Bol, B. & Campbell, V. (2004). A beta-mediated activation of the apoptotic cascade in cultured cortical neurones: a role for cathepsin-L. *Neurobiol Aging,* 25, 83–91.

Borner, C. (2003). The Bcl-2 protein family: sensors and checkpoints for life-or-death decisions. *Mol Immunol. 39(11)*, 615-647.

Boya, P., Andreau, K., Poncet, D., Zamzami, N., Perfettini, J. L., Metivier, D., Ojcius, D. M., Jaattela, M. & Kroemer, G. (2003a). Lysosomal membrane permeabilization induces cell death in a mitochondrion-dependent fashion. *J Exp Med. 197*, 1323-1334.

Boya, P., Gonzalez-Polo, R. A., Poncet, D., Andreau, K., Vieira, H. L., Roumier, T., Perfettini, J. L. & Kroemer, G. (2003b). Mitochondrial membrane permeabilization is a critical step of lysosome-initiated apoptosis induced by hydroxychloroquine. *Oncogene, 22*, 3927-3936.

Bremer, C., Tung, C. H., Bogdanov Jr, A. & Weissleder, R. (2002). Imaging of differential protease expression in breast cancers for detection of aggressive tumor phenotypes. *Radiology, 222*, 814-818.

Bröker, L. E., Huisman, C., Span, S. W., Rodriguez, J. A., Kruyt, F. A. & Giaccone, G. (2004). Cathepsin B mediates caspase-independent cell death induced by microtubule stabilizing agents in non-small cell lung cancer cells. *Cancer Res, 64,* 27-30.

Brunk, U. T. & Svensson, I. (1999). Oxidative stress, growth factor starvation and Fas activation may all cause apoptosis through lysosomal leak. *Redox Rep. 4,* 3-11.

Brunk, U. T., Dalen, H., Roberg, K. & Hellquist, H. B. (1997). Photo-oxidative disruption of lysosomal membranes causes apoptosis of cultured human fibroblasts. *Free Radic Biol Med. 23*, 616-626.

Brunk, U. T., Neuzil, J. & Eaton, J. W. (2001). Lysosomal involvement in apoptosis. *Redox Rep. 6*, 91-97.

Bursch, W. (2001). The autophagosomal-lysosomal compartment in programmed cell death. *Cell Death Differ. 8*, 569-581.

Castino, R., Demoz, M. & Isidoro, C. (2003). Destination 'lysosome': a target organelle for tumour cell killing?. *J Mol Recognit. 16(5)*, 337-348.

Castino, R., Pace, D., Demoz, M., Gargiulo, M., Ariatta, C., Raiteri, E. & Isidoro, C. (2002). Lysosomal proteases as potential targets for the induction of apoptotic cell death in human neuroblastomas. *Int J Cancer. 97*, 775-779.

Chipuk, J. E. & Green, D. R. (2005). Do inducers of apoptosis trigger caspase-independent cell death?. *Nat Rev Mol Cell Biol. 6(3)*, 268-275.

Chwieralski, C. E., Welte, T. & Buhling, F. (2006). Cathepsin-regulated apoptosis. *Apoptosis, 11(2)*, 143-149.

Cirman, T., Orešić, K., Mazovec, G. D., Turk, V., Reed, J. C., Myers, R. M., Salvesen, G. S. & Turk, B. (2004). Selective disruption of lysosomes in HeLa cells triggers apoptosis mediated by cleavage of Bid by multiple papain-like lysosomal cathepsins. *J Biol Chem. 279*, 3578-3587.

de Duve, C. (1983). Lysosomes revisited. *Eur J Biochem. 137*, 391-397.

Debnath, J., Baehrecke, E. H. & Kroemer, G. (2005). Does autophagy contribute to cell death?. *Autophagy, 2*, 10-18.

Deiss, L. P., Galinka, H., Berissi, H., Cohen, O. & Kimchi, A. (1996). Cathepsin D protease mediates programmed cell death induced by interferon-gamma, Fas/APO-1 and TNF-alpha. *EMBO J. 15*, 3861-3870.

Demchik, L. L., Sameni, M., Nelson, K., Mikkelsen, T. & Sloane, B. F. (1999). Cathepsin B and glioma invasion. *Int J Dev Neurosci. 17(5-6)*, 483-494.

Demoz, M., Castino, R., Cesaro, P., Baccino, F. M., Bonelli, G. & Isidoro, C. (2002). Endosomal-lysosomal proteolysis mediates death signalling by TNFalpha, not by etoposide, in L929 fibrosarcoma cells: evidence for an active role of cathepsin D. *Biol Chem. 383(7-8)*, 1237-1248.

Desagher, S, & Martinou, J. C. (2000). Mitochondria as the central control point of apoptosis. *Trends Cell Biol. 10(9)*, 369-77.

Desagher, S., Osen-Sand, A., Nichols, A., Eskes, R., Montessuit, S., Lauper, S., Maundrell, K., Antonsson, B. & Martinou, J. C. (1999). Bid-induced conformational change of Bax is responsible for mitochondrial cytochrome c release during apoptosis. *J Cell Biol. 144*, 891-901.

Emert-Sedlak, L., Shangary, S., Rabinovitz, A., Miranda, M. B., Delach, S. M. & Johnson, D. E. (2005). Involvement of cathepsin D in chemotherapy-induced cytochrome *c* release, caspase activation, and cell death. *Mol Cancer Ther. 4*, 733-742.

Erdal, H., Berndtsson, M., Castro, J., Brunk, U., Shoshan, M. C. & Linder, S. (2005). Induction of lysosomal membrane permeabilization by compounds that activate p53-independent apoptosis. *Proc Natl Acad Sci USA, 102*, 192-197.

Eskes, R., Desagher, S., Antonsson, B. & Martinou, J. C. (2000). Bid induces the oligomerization and insertion of Bax into the outermitochondrial membrane. *Mol Cell Biol. 20*, 929-935.

Faubion, W. A., Guicciardi, M .E., Miyoshi, H., Bronk, S. F., Roberts, P. J., Svingen, P. A., Kaufmann, S. H. & Gores, G. J. (1999). Toxic bile salts induce rodent hepatocyte apoptosis via direct activation of Fas. *J Clin Invest. 103*, 137-145.

Fedorowski, A., Steciwko, A. & Rabczynski, J. (2004). Serum cathepsin B activity during regression of Morris hepatoma 5123 D. *Med Sci Monit. 10*, BR144-BR150.

Fehrenbacher, N. & Jaattela, M. (2005). Lysosomes as targets for cancer therapy. *Cancer Res. 65*, 2993-2995.

Fehrenbacher, N., Gyrd-Hansen, M., Poulsen, B., Felbor, U., Kallunki, T., Boes, M., Weber, E., Leist, M. & Jaattela, M. (2004). Sensitization to the lysosomal cell death pathway upon immortalization and transformation. *Cancer Res. 64*, 5301-5310.

Ferri, K. F. & Kroemer, G. (2001). Organelle-specific initiation of cell death pathways. *Nat Cell Biol. 3*, E255-E263.

Fisher, D. E. (1994). Apoptosis in cancer therapy: crossing the threshold. *Cell, 78*, 539-542.

Foghsgaard, L., Wissing, D., Mauch, D., Lademann, U., Bastholm, L., Boes, M., Elling, F., Leist, M. & Jäättelä, M. (2001). Cathepsin B acts as a dominant execution protease in tumor cell apoptosis induced by tumor necrosis factor. *J Cell Biol. 153*, 999-1010.

Friesen, C., Fulda, S. & Debatin, K. M. (1999). Cytotoxic drugs and the CD95 pathway. *Leukemia, 13*, 1854-1858.

Gondi, C. S., Lakka, S. S., Dinh, D. H., Olivero, W. C., Gujrati, M. & Rao, J. S. (2004). RNAi-mediated inhibition of cathepsin B and uPAR leads to decreased cell invasion, angiogenesis and tumor growth in gliomas. *Oncogene, 23(52)*, 8486-8496.

Green, D. R. (2000). Apoptotic pathways: paper wraps stone blunts scissors. *Cell, 102*, 1-4.

Guicciardi, M. E., Deussing, J., Miyoshi, H., Bronk, S. F., Svingen, P. A, Peters, C., Kaufmann, S. H. & Gores, G. J. (2000) Cathepsin B contributes to TNF-alpha-mediated hepatocyte apoptosis by promoting mitochondrial release of cytochrome c. *J Clin Invest. 106*, 1127-1137.

Guicciardi, M. E., Leist, M. & Gores, G. J. (2004). Lysosomes in cell death. *Oncogene, 23*, 2881-2890.

Gurbuxani, S., Bruey, J. M., Fromentin, A., Larmonier, N., Parcellier, A., Jäättelä, M., Martin, F., Solary, E. & Garrido, C. (2001). Selective depletion of inducible HSP70 enhances immunogenicity of rat colon cancer cells. *Oncogene, 20*, 7478-7485.

Halestrap, A. P., Gillespie, J. P., O'Toole, A. & Doran, E. (2000). Mitochondria and cell death: a pore way to die?. *Symp Soc Exp Biol. 52*, 65-80.

Hasilik, A. & Neufeld, E. F. (1980). Biosynthesis of lysosomal enzymes in fibroblasts. Phosphorylation of mannose residues. *J Biol Chem. 255*, 4946-4950.

Heinrich, M., Neumeyer, J., Jakob, M., Hallas, C., Tchikov, V., Winoto-Morbach, S., Wickel, M., Schneider-Brachert, W., Trauzold, A., Hethke, A. & Schutze, S. (2004). Cathepsin D links TNF-induced acid sphingomyelinase to Bid-mediated caspase-9 and -3 activation. *Cell Death Differ. 11*, 550-563.

Hengartner, M. O. (2000). The biochemistry of apoptosis. *Nature, 407*, 770-776.

Herr, I. & Debatin, K. M. (2001). Cellular stress response and apoptosis in cancer therapy. *Blood, 98(9)*, 2603-2614.

Hirpara, J. L., Clement, M. V. & Pervaiz, S. (2001). Intracellular acidification triggered by mitochondrial-derived hydrogen peroxide is an effector mechanism for drug-induced apoptosis in tumor cells. *J Biol Chem. 276*, 514-521.

Hishita, T., Tada-Oikawa, S., Tohyama, K., Miura, Y., Nishihara, T., Tohyama, Y., Yoshida, Y., Uchiyama, T. & Kawanishi, S. (2001). Caspase-3 activation by lysosomal enzymes in cytochrome c-independent apoptosis in myelodysplastic syndrome-derived cell line P39. *Cancer Res. 61(7)*, 2878-2884.

Isahara, K., Ohsawa, Y., Kanamori, S., Shibata, M., Waguri, S., Sato, N., Gotow, T., Watanabe, T., Momoi, T., Urase, K., Kominami, E. & Uchiyama, Y. (1999). Regulation of a novel pathway for cell death by lysosomal aspartic and cysteine proteinases. *Neuroscience, 91*, 233-249.

Ishidoh, K. & Kominami, E. (2002). Processing and activation of lysosomal proteinases. *Biol Chem. 383(12)*, 1827-1831.

Jäättelä, M. (1999). Heat shock proteins as cellular lifeguards. *Ann Med. 31*, 261-271.

Jaattela, M. (2002). Programmed cell death: many ways for cells to die decently. *Ann Med. 34*, 480-488.

Jäättelä, M. (2004). Multiple cell death pathways as regulators of tumour initiation and progression. *Oncogene, 23*, 2746-2756.

Jäättelä, M., Wissing, D., Kokholm, K., Kallunki, T. & Egeblad, M. (1998). Hsp70 exerts its anti-apoptotic function downstream of caspase-3-like proteases. *EMBO J. 17*, 6124-6134.

Johansson, A. C., Steen, H., Ollinger, K. & Roberg, K. (2003). Cathepsin D mediates cytochrome c release and caspase activation in human fibroblast apoptosis induced by staurosporine. *Cell Death Differ. 10(11)*, 1253-1259.

Jones, B., Roberts, P. J., Faubion, W. A., Kominami, E. & Gores, G. J. (1998). Cystatin A expression reduces bile salt-induced apoptosis in a rat hepatoma cell line. *Am. J. Physiol. 275*, G723-G730.

Journet, A. & Ferro, M. (2004). The potentials of MS-based subproteomic approaches in medical science: the case of lysosomes and breast cancer. *Mass Spectrom Rev. 23*, 393-442.

Joyce, J. A. & Hanahan, D. (2004). Multiple roles for cysteine cathepsins in cancer. *Cell Cycle, 3*, 1516-1619.

Kagedal, K., Johansson, U. & Ollinger, K. (2001a). The lysosomal protease cathepsin D mediates apoptosis induced by oxidative stress. *FASEB J. 15(9)*, 1592-1594.

Kagedal, K., Zhao, M., Svensson, I. & Brunk, U. T. (2001b). Sphingosine-induced apoptosis is dependent on lysosomal proteases. *Biochem J. 359*, 335-343.

Kang, J. X., Bell, J., Beard, R. L. & Chandraratna, R. A. (1999). Mannose 6-phosphate/insulin-like growth factor II receptor mediates the growth-inhibitory effects of retinoids. *Cell Growth Differ. 10,* 591-600.

Katunuma, N., Matsui, A., Le, Q. T., Utsumi, K., Salvesen, G. & Ohashi, A. (2001). Novel procaspase-3 activating cascade mediated by lysoapoptases and its biological significances in apoptosis. *Adv Enzyme Regul. 41*, 237-250.

Katunuma, N., Murata, E., Le, Q. T., Hayashi, Y. & Ohashi, A. (2004). New apoptosis cascade mediated by lysosomal enzyme and its protection by epigallo-catechin gallate. *Adv Enzyme Regul. 44*, 1 – 10.

Kielan, W., Suzanowicz, J., SiewiNski, M., Saleh, Y., Janocha, A., Skalski, A. & Tarnawa, R. (2004). Evaluation of changes in the activity of proteolytic enzymes and their inhibitors in the processes that accompany the growth of gastric cancer. *Gastric Cancer, 7*, 17-23.

Kirschke, H. (1997). Lysosomal cysteine peptidases and malignant tumours. *Adv Exp Med Biol.* 421, 253-257.

Koblinski, J. E., Ahram, M. & Sloane, B. F. (2000). Unraveling the role of proteases in cancer. *Clin Chim Acta, 291*, 113-135.

Kos, J. & Lah, T. T. (1998). Cysteine proteinases and their endogenous inhibitors: target proteins for prognosis, diagnosis and therapy in cancer (review). *Oncol Rep. 5*, 1349-1361.

Kos, J., Werle, B., Lah, T. & Brunner, N. (2000). Cysteine proteinases and their inhibitors in extracellular fluids: markers for diagnosis and prognosis in cancer. *Int J Biol Markers, 15*, 84-89.

Koster, A., Saftig, P., Matzner, U., von Figura, K., Peters, C. & Pohlmann, R. (1993). Targeted disruption of the M(r) 46,000 mannose 6-phosphate receptor gene in mice results in misrouting of lysosomal proteins. *EMBO J. 12*, 5219-5223.

Krammer, P. H. (2000). CD95's deadly mission in the immune system. *Nature, 407*, 789-795.

Kroemer, G. & Jaattela, M. (2005). Lysosomes and autophagy in cell death control. *Nat Rev Cancer, 5*, 886-897.

Kuma, A., Hatano, M., Matsui, M., Yamamoto, A., Nakaya, H., Yoshimori, T., Ohsumi, Y., Tokuhisa, T. & Mizushima, N. (2004). The role of autophagy during the early neonatal starvation period. *Nature, 432*, 1032-1036.

Kuopio, T., Kankaanranta, A., Jalava, P., Kronqvist, .P, Kotkansalo, T., Weber, E. & Collan, Y. (1998). Cysteine proteinase inhibitor cystatin A in breast cancer. *Cancer Res. 58*, 432-436.

Lazaris, A., Chatzigianni, E. B., Panoussopoulos, D., Tzimas, G. N., Davaris, P. S. & Golematis, B. (1997). Proliferating cell nuclear antigen and heat shock protein 70 immunolocalization in invasive ductal breast cancer not otherwise specified. *Breast Cancer Res Treat. 43*, 43-51.

Leist, M. & Jäättela, M. (2001). Four deaths and a funeral: from caspases to alternative mechanisms. *Nat Rev Mol Cell Biol. 2*, 589-598.

Levicar, N., Dewey, R. A., Daley, E., Bates, T. E., Davies, D., Kos, J., Pilkington, G. J. & Lah, T. T. (2003). Selective suppression of cathepsin L by antisense cDNA impairs human brain tumor cell invasion in vitro and promotes apoptosis. *Cancer Gene Ther. 10*, 141-151.

Levine, B. & Yuan, J. (2005). Autophagy in cell death: an innocent convict?. *J Clin Invest. 115*, 2679–2688.

Li, H., Zhu, H., Xu, C. J. & Yuan, J. (1998). Cleavage of BID by caspase 8 mediates the mitochondrial damage in the Fas pathway of apoptosis. *Cell, 94*, 491-501.

Li, W., Ding, F., Zhang, L., Liu, Z., Wu, Y., Luo, A., Wu, M., Wang, M., Zhan, Q. & Liu, Z. (2005). Overexpression of stefin A in human esophageal squamous cell carcinoma cells inhibits tumor cell growth, angiogenesis, invasion, and metastasis. *Clin Cancer Res. 11(24 Pt 1)*, 8753-8762.

Li, W., Yuan, X., Nordgren, G., Dalen, H., Dubowchik, G. M., Firestone, R. A. & Brunk, U. T. (2000) Induction of cell death by the lysosomotropic detergent MSDH. *FEBS Lett. 470*, 35-39.

Li, X., Rayford, H., Shu, R., Zhuang, J. & Uhal, B. D. (2004). Essential role for cathepsin D in bleomycin-induced apoptosis of alveolar epithelial cells. *Am J Physiol: Lung Cell Mol Physiol. 287*, L46-L51.

Liu, N., Raja, S. M., Zazzeroni, F., Metkar, S. S., Shah, R., Zhang, M., Wang, Y., Brömme, D., Russin, W. A., Lee, J. C., Peter, M. E., Froelich, C. J., Franzoso, G. & Ashton-Rickardt, P. G. (2003). NF-kappaB protects from the lysosomal pathway of cell death. *EMBO J. 22*, 5313-5322.

Liu, N., Wang, Y. & Ashton-Rickardt, P. G. (2004). Serine protease inhibitor 2A inhibits caspase-independent cell death. *FEBS Lett. 569*, 49-53.

Lobel, P., Fujimoto, K., Ye, R. D., Griffiths, G. & Kornfeld, S. (1989). Mutations in the cytoplasmic domain of the 275 kd mannose 6-phosphate receptor differentially alter lysosomal enzyme sorting and endocytosis. *Cell, 57*, 787-796.

Lockshin, R. A. & Zakeri, Z. (2004). Apoptosis, autophagy, and more. *Int. J. Biochem. Cell Biol. 36*, 2405-2419.

Losch, A., Schindl, M., Kohlberger, P., Lahodny, J., Breitenecker, G., Horvat, R. & Birner, P. (2004). Cathepsin D in ovarian cancer: prognostic value and correlation with p53 expression and microvessel density. *Gynecol Oncol. 92(2),* 545-552.

Lum, J. J., Bauer, D. E., Kong, M., Harris, M. H., Li, C., Lindsten, T. & Thompson, C. B. (2005). Growth factor regulation of autophagy and cell survival in the absence of apoptosis. *Cell, 120*, 237-248.

Madge, L. A., Li, J. H., Choi, J. & Pober, J. S. (2003). Inhibition of phosphatidylinositol 3-kinase sensitizes vascular endothelial cells to cytokine-initiated cathepsin-dependent apoptosis. *J Biol Chem. 278*, 21295-21306.

Majeski, A. E. & Dice, J. F. (2004). Mechanisms of chaperone-mediated autophagy. *Int J Biochem Cell Biol. 36*, 2435-2444.

Mathiasen, I. S., Hansen, C. M., Foghsgaard, L. & Jaattela, M. (2001). Sensitization to TNF-induced apoptosis by 1, 25-dihydroxy vitamin D (3) involves up-regulation of the TNF receptor 1 and cathepsin B. *Int J Cancer, 93*, 224-231.

Miller, D. K., Griffiths, E., Lenard, J. & Firestone, R. A. (1983). Cell killing by lysosomotropic detergents. *J Cell Biol. 97*, 1841-1851.

Mortimore, G. E. & Kadowaki, M. *Cellular Proteolytic Systems*. In: Ciechanover, A. J. & Schwartz, A. L., editors. New York, NY: Wiley-Liss; 1994; 65-87.

Mosser, D. D., Caron, A. W., Bourget, L., Denis-Larose, C. & Massie, B. (1997). Role of the human heat shock protein hsp70 in protection against stress-induced apoptosis. *Mol Cell Biol. 17*, 5317-5327.

Mousavi, S. A., Brech, A., Berg, T. & Kjeken, R. (2003). Phosphoinositide 3-kinase regulates maturation of lysosomes in rat hepatocytes. *Biochem J. 372*, 861-869.

Nagaraj, N. S., Vigneswaran, N. & Zacharias, W. (2004). Hypoxia-mediated apoptosis in oral carcinoma cells occurs via two independent pathways. *Mol Cancer, 3(1)*, 38.

Nagaraj, N. S., Vigneswaran, N. & Zacharias, W. (2006). Cathepsin B mediates TRAIL-induced apoptosis in oral cancer cells. *J Cancer Res Clin Oncol. 132(3)*, 171-183.

Nomura, T. & Katunuma, N. (2005). Involvement of cathepsins in the invasion, metastasis and proliferation of cancer cells. *J Med Invest. 52*, 1-9.

Nylandsted, J., Gyrd-Hansen, M., Danielewicz, A., Fehrenbacher, N., Lademann, U., Hoyer-Hansen, M., Weber, E., Multhoff, G., Rohde, M. & Jaattela, M. (2004). Heat shock protein 70 promotes cell survival by inhibiting lysosomal membrane permeabilization. *J Exp Med. 200(4)*, 425-35.

Ollinger, K. (2000). Inhibition of cathepsin D prevents free-radical-induced apoptosis in rat cardiomyocytes. *Arch Biochem Biophys. 373*, 346-351.

Ono, K., Kim, S. O. & Han, J. (2003). Susceptibility of lysosomes to rupture is a determinant for plasma membrane disruption in tumor necrosis factor alpha-induced cell death. *Mol Cell Biol. 23(2)*, 665-676.

Persson, H. L., Yu, Z., Tirosh, O., Eaton, J. W. & Brunk, U. T. (2003). Prevention of oxidant-induced cell death by lysosomotropic iron chelators. *Free Radical Biol Med. 34*, 1295-1305.

Reid, W. A., Valler, M. J. & Kay, J. (1986). Immunolocalization of cathepsin D in normal and neoplastic human tissues. *J Clin Pathol. 39*, 1323-1330.

Reiners Jr, J. J., Caruso, J. A., Mathieu, P., Chelladurai, B., Yin, X. M. & Kessel, D. (2002). Release of cytochrome *c* and activation of pro-caspase-9 following lysosomal photodamage involves Bid cleavage. *Cell Death Differ. 9*, 934-944.

Roberg, K. & Ollinger, K. (1998). Oxidative stress causes relocation of the lysosomal enzyme cathepsin D with ensuing apoptosis in neonatal rat cardiomyocytes. *Am J Pathol. 152*, 1151-1156.

Roberg, K., Johansson, U. & Ollinger, K. (1999). Lysosomal release of cathepsin D precedes relocation of cytochrome *c* and loss of mitochondrial transmembrane potential during apoptosis induced by oxidative stress. *Free Radic Biol Med. 27*, 1228-1237.

Roberg, K., Kagedal, K. & Ollinger, K. (2002). Microinjection of cathepsin d induces caspase-dependent apoptosis in fibroblasts. *Am J Pathol. 161*, 89-96.

Roberts, L. R., Adjei, P. N. & Gores, G. J. (1999). Cathepsins as effector proteases in hepatocyte apoptosis. *Cell Biochem Biophys. 30*, 71-88.

Roberts, L. R., Kurosawa, H., Bronk, S. F., Fesmier, P. J., Agellon, L. B., Leung, W. Y., Mao, F. & Gores, G. J. (1997). Cathepsin B contributes to bile salt-induced apoptosis of rat hepatocytes. *Gastroenterology, 113*, 1714-1726.

Rochefort, H., Garcia, M., Glondu, M., Laurent, V., Liaudet, E., Rey, J. M. & Roger, P. (2000). Cathepsin D in breast cancer: mechanisms and clinical applications, a 1999 overview. *Clin Chim Acta. 291(2)*, 157-170.

Schmitt, C. A. & Lowe, S. W. (2001). Apoptosis is critical for drug response *in vivo. Drug Resist Updat. 4*, 132-144.

Schwartz, M. K. (1995). Tissue cathepsins as tumor markers. *Clin Chim Acta. 237*, 67-78.

Shibata, M., Kanamori, S., Isahara, K., Ohsawa, Y., Konishi, A., Kametaka, S., Watanabe, T., Ebisu, S., Ishido, K., Kominami, E. & Uchiyama, Y. (1998). Participation of cathepsins B and D in apoptosis of PC12 cells following serum deprivation. *Biochem Biophys Res Commun. 251*, 199-203.

Simon, M. M., Krone, C., Schwarz, A., Luger, T. A., Jäättelä, M. & Schwarz, T. (1995). Heat shock protein 70 overexpression affects the response to ultraviolet light in murine fibroblasts. Evidence for increased cell viability and suppression of cytokine release. *J Clin Invest. 95*, 926-933.

Stein, M., Zijderhand-Bleekemolen, J. E., Geuze, H., Hasilik, A. & von Figura, K. (1987). Mr 46,000 mannose 6-phosphate specific receptor: its role in targeting of lysosomal enzymes. *EMBO J. 6*, 2677-2681.

Stoka, V., Turk, B., Schendel, S. L., Kim, T. H., Cirman, T., Snipas, S. J., Ellerby, L. M., Bredesen, D., Freeze, H., Abrahamson, M., Brömme, D., Krajewski, S., Reed, J. C., Yin, X. M., Turk, V. & Salvesen, G. S. (2001). Lysosomal protease pathways to apoptosis. Cleavage of bid, not pro-caspases, is the most likely route. *J Biol Chem. 276*, 3149-3157.

Taha, T. A., Kitatani, K., Bielawski, J., Cho, W., Hannun, Y. A. & Obeid, L. M. (2005). Tumor necrosis factor induces the loss of sphingosine kinase-1 by a cathepsin B dependent mechanism. *J Biol Chem. 280*, 17196-17202.

Taha, T. A., Kitatani, K., Bielawski, J., Cho, W., Hannun, Y. A. & Obeid, L. M. (2004). Down-regulation of sphingosine kinase-1 by DNA damage: dependence on proteases and p53. *J Biol Chem. 279,* 20546-20554.

Tang, J. & Wong, R. N. (1987). Evolution in the structure and function of aspartic proteases. *J Cell Biochem. 33*, 53-63.

Tardy, C., Autefage, H., Garcia, V., Levade, T. & Andrieu-Abadie, N. (2004). Mannose 6-phosphorylated proteins pre pequired for pumor necrosis factor-induced apoptosis: defective response in I-cell disease fibroblasts. *J. Biol. Chem. 279*, 52914-52923.

Tardy, C., Codogno, P., Autefage, H., Levade, T. & Andrieu-Abadie, N. (2006). Lysosomes and lysosomal proteins in cancer cell death (new players of an old struggle). *Biochim Biophys Acta. 1765(2)*, 101-125.

Tardy, C., Tyynela, J., Hasilik, A., Levade, T. & Andrieu-Abadie, N. (2003). Stress-induced apoptosis is impaired in cells with a lysosomal targeting defect but is not affected in cells synthesizing a catalytically inactive cathepsin D. *Cell Death Differ. 10*, 1090-1100.

Thibodeau, M. S., Giardina, C., Knecht, D. A., Helble, J. & Hubbard, A. K. (2004). Silica-induced apoptosis in mouse alveolar macrophages is initiated by lysosomal enzyme activity. *Toxicol Sci. 80*, 34-48.

Thomssen, C., Schmitt, M., Goretzki, L., Oppelt, P., Pache, L., Dettmar, P., Janicke, F. & Graeff, H. (1995). Prognostic value of the cysteine proteases cathepsins B and cathepsin L in human breast cancer. *Clin Cancer Res. 1*, 741-746.

Tobin, D., Arvanitidis, M. & Bisby, R. H. (2002). One-electron oxidation of "photo-Fenton" reagents and inhibition of lipid peroxidation. *Biochem Biophys Res Commun. 299(1)*, 155-159.

Tsuchiya, K., Kohda, Y., Yoshida, M., Zhao, L., Ueno, T., Yamashita, J., Yoshioka, T., Kominami, E. & Yamashima, T. (1999). Postictal blockade of ischemic hippocampal neuronal death in primates using selective cathepsin inhibitors. *Exp Neurol. 155*, 187-194.

Turk, B., Turk, D. & Salvesen, G. S. (2002). Regulating cysteine protease activity: essential role of protease inhibitors as guardians and regulators. *Curr Pharm Des. 8*, 1623-1637.

Turk, B., Turk, D. & Turk, V. (2000). Lysosomal cysteine proteases: more than scavengers. *Biochim Biophys Acta. 1477*, 98-111.

Turk, B., Turk, V. & Turk, D. (1997). Structural and functional aspects of papain-like cysteine proteinases and their protein inhibitors. *Biol Chem. 378*, 141-150.

Turk, V. & Bode, W. (1991). The cystatins: protein inhibitors of cysteine proteinases. *FEBS Lett. 285*, 213-219.

Vancompernolle, K., Van Herreweghe, F., Pynaert, G., Van de Craen, M., De Vos, K., Totty, N., Sterling, A., Fiers, W., Vandenabeele, P. & Grooten, J. (1998). Atractyloside-induced release of cathepsin B, a protease with caspase-processing activity. *FEBS Lett. 438(3)*, 150-158.

Vargas-Roig, L. M., Gago, F. E., Tello, O., Aznar, J. C. & Ciocca, D. R. (1998). Heat shock protein expression and drug resistance in breast cancer patients treated with induction chemotherapy. *Intl J Cancer, 79*, 468-475.

Vayssier, M., Banzet, N., Francois, D., Bellmann, K. & Polla, B. S. (1998). Tobacco smoke induces both apoptosis and necrosis in mammalian cells: Differential effects of HSP70. *Am J Physiol. 275*, L771-L779.

Vigneswaran, N., Wu, J., Nagaraj, N., Adler-Storthz, K. & Zacharias, W. (2005). Differential susceptibility of metastatic and primary oral cancer cells to TRAIL-induced apoptosis. *Int J Oncol. 26(1)*, 103-112.

Vigneswaran, N., Wu, J., Nagaraj, N., James, R., Zeeuwen, P. & Zacharias, W. (2006). Silencing of cystatin M in metastatic oral cancer cell line MDA-686Ln by siRNA increases cysteine proteinases and legumain activities, cell proliferation and in vitro invasion. *Life Sci. 78(8)*, 898-907.

Vigneswaran, N., Zhao, W., Dassanayake, A., Muller, S., Miller, D. M. & Zacharias, W. (2000). Variable expression of cathepsin B and D correlates with highly invasive and metastatic phenotype of oral cancer. *Hum Pathol. 31(8)*, 931-937.

Visscher, D. W., Sarkar, F., LoRusso, P., Sakr, W., Ottosen, S., Wykes, S. & Crissman, J. D. (1993). Immunohistologic evaluation of invasion-associated proteases in breast carcinoma. *Mod Pathol. 6*, 302-306.

Volloch, V. Z. & Sherman, M. Y. (1999). Oncogenic potential of Hsp72. *Oncogene, 18*, 3648-3651.

von Figura, K. & Hasilik, A. (1986). Lysosomal enzymes and their receptors?. *Annu Rev Biochem. 55*, 167-193.

Walczak, H., Miller, R. E., Ariail, K., Gliniak, B., Griffith, T. S., Kubin, M., Chin, W., Jones, J., Woodward, A., Le, T., Smith, C., Smolak, P., Goodwin, R. G., Rauch, C. T., Schuh, J. C. & Lynch, D. H. (1999). Tumoricidal activity of tumor necrosis factor-related apoptosis-inducing ligand in vivo. *Nat Med. 5(2)*, 157-163.

Welss, T., Sun, J., Irving, J. A., Blum, R., Smith, A. I., Whisstock, J. C., Pike, R. N., von Mikecz, A., Ruzicka, T., Bird, P. I. & Abts, H. F. (2003). Hurpin is a selective inhibitor of lysosomal cathepsin L and protects keratinocytes from ultraviolet-induced apoptosis. *Biochemistry, 42(24)*, 7381-7389.

Werneburg, N. W., Guicciardi, M. E., Bronk, S. F. & Gores, G. J. (2002). Tumor necrosis factor-alpha-associated lysosomal permeabilization is cathepsin B dependent. *Am J Physiol Gastrointest Liver Physiol. 283*, G947–G956.

Wex, T., Levy, B., Wex, H. & Bromme, D. (1999). Human cathepsins F and W: A new subgroup of cathepsins. *Biochem Biophys Res Communications, 259*, 401-407.

Wickramasinghe, N. S., Nagaraj, N. S., Vigneswaran, N. & Zacharias, W. (2005). Cathepsin B promotes both motility and invasiveness of oral carcinoma cells. *Arch Biochem Biophys. 436(1)*, 187-195.

Willst¨atter, R. & Bamann, E. (1929). ¨Uber die Proteasen der Magenschleinhaut. *Hoppe-Seyler's Z Physiol Chem. 180*, 127-143.

Wu, G. S., Saftig, P., Peters, C. & El-Deiry, W. S. (1998). Potential role for cathepsin D in p53-dependent tumor suppression and chemosensitivity. *Oncogene, 16*, 2177-2183.

Yamashima, T. (2000). Implication of cysteine proteases calpain, cathepsin and caspase in ischemic neuronal death of primates. *Prog Neurobiol. 62(3)*, 273-295.

Yamashima, T., Kohda, Y., Tsuchiya, K., Ueno, T., Yamashita, J., Yoshioka, T. & Kominami, E. (1998). Inhibition of ischaemic hippocampal neuronal death in primates with cathepsin B inhibitor CA-074: a novel strategy for neuroprotection based on 'calpain-cathepsin hypothesis'. *Eur J Neurosci. 10*, 1723-1733.

Yin, L., Stearns, R. & Gonzalez-Flecha, B. (2005). Lysosomal and mitochondrial pathways in H_2O_2-induced apoptosis of alveolar type II cells. *J Cell Biochem. 94*, 433-445.

Yuan, X. M., Li, W., Dalen, H., Lotem, J., Kama, R., Sachs, L. & Brunk, U. T. (2002). Lysosomal destabilization in p53-induced apoptosis. *Proc Natl Acad Sci USA, 99*, 6286-6291.

Zang, Y., Beard, R. L., Chandraratna, R. A. & Kang, J. X. (2001). Evidence of a lysosomal pathway for apoptosis induced by the synthetic retinoid CD437 in human leukemia HL-60 cells. *Cell Death Differ. 8*, 477-485.

Zhao, M., Antunes, F., Eaton, J. W. & Brunk, U. T. (2003). Lysosomal enzymes promote mitochondrial oxidant production, cytochrome c release and apoptosis. *Eur J Biochem. 270*, 3778-3786.

Zhao, M., Eaton, J. W. & Brunk, U. T. (2001). Bcl-2 phosphorylation is required for inhibition of oxidative stress-induced lysosomal leak and ensuing apoptosis. *FEBS Lett. 509(3)*, 405-412.

INDEX

B

C

D

E

F

G

H

I

J

K

L

M

N

R

S

T

U

V

W

X

Y

Z